MATHEMATICS HIGHER LEVEL: STATISTICS AND PROBABILITY

COURSE COMPANION

Josip Harcet
Lorraine Heinrichs
Palmira Mariz Seiler
Marlene Torres-Skoumal

OXFORD
UNIVERSITY PRESS

OXFORD
UNIVERSITY PRESS

Great Clarendon Street, Oxford OX2 6DP

Oxford University Press is a department of the University of Oxford.
It furthers the University's objective of excellence in research, scholarship,
and education by publishing worldwide in

Oxford   New York

Auckland Cape Town Dar es Salaam Hong Kong Karachi
Kuala Lumpur Madrid Melbourne Mexico City Nairobi
New Delhi Shanghai Taipei Toronto

With offices in

Argentina Austria Brazil Chile Czech Republic France Greece
Guatemala Hungary Italy Japan South Korea Poland Portugal
Singapore Switzerland Thailand Turkey Ukraine Vietnam

Oxford is a registered trade mark of Oxford University Press
in the UK and in certain other countries

British Library Cataloguing in Publication Data

Data available

ISBN 978-0-19-830485-2
10 9 8 7 6 5

Printed in Great Britain

Paper used in the production of this book is a natural, recyclable product made from wood grown in sustainable forests.
The manufacturing process conforms to the environmental regulations of the country of origin.

Acknowledgements

The publisher would like to thank the following for permission to reproduce photographs:

p4: Nathaniel S. Butler/NBA/Getty Images; **p12:** Kuttig-People/Alamy; **p41:** Jaggat Rashidi/Shutterstock; **p79:** Ariwasabi/
Shutterstock; **p84:** Anton Havelaar/Shutterstock; **p123:** evantravels/Shutterstock; **p128:** NASA.

Course Companion definition

The IB Diploma Programme Course Companions are resource materials designed to support students throughout their two-year Diploma Programme course of study in a particular subject. They will help students gain an understanding of what is expected from the study of an IB Diploma Programme subject while presenting content in a way that illustrates the purpose and aims of the IB. They reflect the philosophy and approach of the IB and encourage a deep understanding of each subject by making connections to wider issues and providing opportunities for critical thinking.

The books mirror the IB philosophy of viewing the curriculum in terms of a whole-course approach; the use of a wide range of resources, international mindedness, the IB learner profile and the IB Diploma Programme core requirements, theory of knowledge, the extended essay, and creativity, action, service (CAS).

Each book can be used in conjunction with other materials and indeed, students of the IB are required and encouraged to draw conclusions from a variety of resources. Suggestions for additional and further reading are given in each book and suggestions for how to extend research are provided.

In addition, the Course Companions provide advice and guidance on the specific course assessment requirements and on academic honesty protocol. They are distinctive and authoritative without being prescriptive.

IB mission statement

The International Baccalaureate aims to develop inquiring, knowledgable and caring young people who help to create a better and more peaceful world through intercultural understanding and respect.

To this end the IB works with schools, governments and international organizations to develop challenging programmes of international education and rigorous assessment.

These programmes encourage students across the world to become active, compassionate, and lifelong learners who understand that other people, with their differences, can also be right.

The IB learner Profile

The aim of all IB programmes is to develop internationally minded people who, recognizing their common humanity and shared guardianship of the planet, help to create a better and more peaceful world. IB learners strive to be:

Inquirers They develop their natural curiosity. They acquire the skills necessary to conduct inquiry and research and show independence in learning. They actively enjoy learning and this love of learning will be sustained throughout their lives.

Knowledgable They explore concepts, ideas, and issues that have local and global significance. In so doing, they acquire in-depth knowledge and develop understanding across a broad and balanced range of disciplines.

Thinkers They exercise initiative in applying thinking skills critically and creatively to recognize and approach complex problems, and make reasoned, ethical decisions.

Communicators They understand and express ideas and information confidently and creatively in more than one language and in a variety of modes of communication. They work effectively and willingly in collaboration with others.

Principled They act with integrity and honesty, with a strong sense of fairness, justice, and respect for the dignity of the individual, groups, and communities.

They take responsibility for their own actions and the consequences that accompany them.

Open-minded They understand and appreciate their own cultures and personal histories, and are open to the perspectives, values, and traditions of other individuals and communities. They are accustomed to seeking and evaluating a range of points of view, and are willing to grow from the experience.

Caring They show empathy, compassion, and respect towards the needs and feelings of others. They have a personal commitment to service, and act to make a positive difference to the lives of others and to the environment.

Risk-takers They approach unfamiliar situations and uncertainty with courage and forethought, and have the independence of spirit to explore new roles, ideas, and strategies. They are brave and articulate in defending their beliefs.

Balanced They understand the importance of intellectual, physical, and emotional balance to achieve personal well-being for themselves and others.

Reflective They give thoughtful consideration to their own learning and experience. They are able to assess and understand their strengths and limitations in order to support their learning and personal development.

A note of academic honesty

It is of vital importance to acknowledge and appropriately credit the owners of information when that information is used in your work. After all, owners of ideas (intellectual property) have property rights. To have an authentic piece of work, it must be based on your individual and original ideas with the work of others fully acknowledged. Therefore, all assignments, written or oral, completed for assessment must use your own language and expression. Where sources are used or referred to, whether in the form of direct quotation or paraphrase, such sources must be appropriately acknowledged.

How do I acknowledge the work of others?

The way that you acknowledge that you have used the ideas of other people is through the use of footnotes and bibliographies.

Footnotes (placed at the bottom of a page) or endnotes (placed at the end of a document) are to be provided when you quote or paraphrase from another document, or closely summarize the information provided in another document. You do not need to provide a footnote for information that is part of a 'body of knowledge'. That is, definitions do not need to be footnoted as they are part of the assumed knowledge.

Bibliographies should include a formal list of the resources that you used in your work. 'Formal' means that you should use one of the several accepted forms of presentation. This usually involves separating the resources that you use into different categories (e.g. books, magazines, newspaper articles, Internet-based resources, CDs and works of art) and providing full information as to how a reader or viewer of your work can find the same information. A bibliography is compulsory in the extended essay.

What constitutes malpractice?

Malpractice is behaviour that results in, or may result in, you or any student gaining an unfair advantage in one or more assessment component. Malpractice includes plagiarism and collusion.

Plagiarism is defined as the representation of the ideas or work of another person as your own. The following are some of the ways to avoid plagiarism:

- Words and ideas of another person used to support one's arguments must be acknowledged.
- Passages that are quoted verbatim must be enclosed within quotation marks and acknowledged.
- CD-ROMs, email messages, web sites on the Internet, and any other electronic media must be treated in the same way as books and journals.
- The sources of all photographs, maps, illustrations, computer programs, data, graphs, audio-visual, and similar material must be acknowledged if they are not your own work.
- Words of art, whether music, film, dance, theatre arts, or visual arts, and where the creative use of a part of a work takes place, must be acknowledged.

Collusion is defined as supporting malpractice by another student. This includes:

- allowing your work to be copied or submitted for assessment by another student
- duplicating work for different assessment components and/or diploma requirements.

Other forms of malpractice include any action that gives you an unfair advantage or affects the results of another student. Examples include, taking unauthorized material into an examination room, misconduct during an examination, and falsifying a CAS record.

About the book

The new syllabus for Mathematics Higher Level Option: Statistics is thoroughly covered in this book. Each chapter is divided into lesson-size sections with the following features:

The Course Companion will guide you through the latest curriculum with full coverage of all topics and the new internal assessment. The emphasis is placed on the development and improved understanding of mathematical concepts and their real life application as well as proficiency in problem solving and critical thinking. The Course Companion denotes questions that would be suitable for examination practice and those where a GDC may be used.

Questions are designed to increase in difficulty, strengthen analytical skills and build confidence through understanding.

Where appropriate the solutions to examples are given in the style of a graphics display calculator.

Mathematics education is a growing, ever changing entity. The contextual, technology integrated approach enables students to become adaptable, lifelong learners.

Note: US spelling has been used, with IB style for mathematical terms.

About the authors

Lorraine Heinrichs has been teaching mathematics for 30 years and IB mathematics for the past 16 years at Bonn International School. She has been the IB DP coordinator since 2002. During this time she has also been workshop leader of the IB; she was also a member of the curriculum review team.

Palmira Mariz Seiler has been teaching mathematics for over 25 years. She joined the IB community in 2001 as a teacher at the Vienna International School and since then has also worked in curriculum review working groups and as a workshop leader. Currently she teaches at Colegio Anglo Colombiano in Bogota, Colombia.

Marlene Torres-Skoumal has taught IB mathematics for over 30 years. During this time, she has enjoyed various roles with the IB, including workshop leader and a member of several curriculum review teams.

Josip Harcet has been involved with and teaching the IB programme since 1992. He has served as a curriculum review member as a workshop leader since 1998.

Contents

Exploring further probability distributions

1

CHAPTER OBJECTIVES:

7.1 Cumulative distribution functions for both discrete and continuous distributions.
Geometric distribution. Negative binomial (Pascal's) distribution.
Probability generating functions for discrete random variables.
Using probability generating functions to find the mean, variance, and
distribution of the sum of n independent random variables.

Before you start

You should know how to:

1 Find the mode, median, mean, and
standard deviation of a discrete random
variable, e.g. the table shows the probability
distribution of a discrete random variable X.

$X = x_i$	1	2	3	4
$P(X = x_i)$	0.3	0.25	0.35	0.1

Mode $(X) = 3$, because $P(X = 3) = 0.35$,
which is the highest probability of the
four values of the random variable.

Median, $m = 2$ since

$P(X \leq 1) = 0.3$ and $P(X \leq 2) = 0.55$

$\mu = E(X) = \sum_{i=1}^{4} x_i p_i$

$= 1 \times 0.3 + 2 \times 0.25 + 3 \times 0.35 + 4 \times 0.1 = 2.25$

$\sigma = \sqrt{\mathrm{Var}(X)} = \sqrt{\sum_{i=1}^{4} x_i^2 p_i - \mu^2}$

$= \sqrt{1^2(0.3) + 2^2(0.25) + 3^2(0.35) + 4^2(0.1) - 0.225^2}$

$= \sqrt{6.05 - 2.25^2} = 0.994$

Skills check:

1 Find the mode, median, mean and
standard deviation of the following
discrete random variables given by:

a

x_i	−1	0	1	2	3	4
p_i	0.3	0.1	0.3	0.1	0.05	0.15

b $P(X = x) = \begin{cases} \dfrac{5-x}{10}, & x = 1, 2, 3, 4 \\ 0, & \text{otherwise} \end{cases}$

2 Find the mode, median, mean and standard deviation of a continuous random variable, e.g. the probability density function of a discrete random variable X is given by the formula

$$f(x) = \begin{cases} \dfrac{1}{2}x, & 0 \le x \le 2 \\ 0, & \text{elsewhere} \end{cases}$$

Mode $(X) = 2$ because it has the maximum point at the end of the interval.

Median, $\displaystyle\int_0^m f(x)\,dx = \dfrac{1}{2} \Rightarrow \dfrac{m^2}{4} = \dfrac{1}{2} \Rightarrow m = \sqrt{2}$

$\mu = E(X) = \displaystyle\int_0^2 xf(x)\,dx = \int_0^2 \dfrac{1}{2}x^2\,dx = \dfrac{4}{3}$

$\sigma = \sqrt{\mathrm{Var}(X)} = \sqrt{\displaystyle\int_0^2 x^2 f(x) - \mu^2}$

$\sqrt{\displaystyle\int_0^2 \dfrac{1}{2}x^3\,dx - \left(\dfrac{4}{3}\right)^2} = \sqrt{2 - \dfrac{16}{9}} = \dfrac{\sqrt{2}}{3}$

3 Find the sum of an infinite geometric series by using the formula

$$u_1 + u_2 + u_3 + \ldots = \dfrac{u_1}{1 - r}, \quad 0 < |r| < 1$$

e.g. the series

$$\dfrac{9}{2} + 3 + 2 + \dfrac{4}{3} + \ldots = \dfrac{\dfrac{9}{2}}{1 - \dfrac{2}{3}} = \dfrac{27}{2}$$

4 Differentiate and integrate composite functions, e.g.

$$f(x) = \dfrac{2}{(3 - 4x)^3}, \quad x \ne \dfrac{3}{4}$$

$$\Rightarrow f'(x) = \dfrac{2 \times (-3) \times (-4)}{(3 - 4x)^4} = \dfrac{24}{(3 - 4x)^4}$$

$$f(x) = \dfrac{2}{(3 - 4x)^3}$$

$$\Rightarrow \int \dfrac{2}{(3 - 4x)^3}\,dx = \dfrac{1}{4(3 - 4x)^2} + c$$

2 Find the mode, median, mean and standard deviation of the continuous random variables defined by the given probability density function:

a $f(x) = \begin{cases} \dfrac{3}{4} - \dfrac{3}{16}x^2, & 0 \le x \le 2 \\ 0, & \text{elsewhere} \end{cases}$

b $f(x) = \begin{cases} \cos(2x), & -\dfrac{\pi}{4} \le x \le \dfrac{\pi}{4} \\ 0, & \text{elsewhere} \end{cases}$

c $f(x) = \begin{cases} \dfrac{6}{x^2}, & 3 \le x \le 6 \\ 0, & \text{elsewhere} \end{cases}$

3 Find the sum of the following infinite geometric series:

a $1 - 0.5 + 0.25 - 0.125 + \ldots$

b $\sqrt{2} + 1 + \dfrac{\sqrt{2}}{2} + \dfrac{1}{2} \ldots$

4 Differentiate and integrate the following composite functions:

a $f(x) = \dfrac{1}{2 - x}, \quad x \ne 2$

b $f(x) = e^{3x+1}$

c $f(x) = \dfrac{3}{2} \sin \dfrac{\pi - 2x}{3}$

d $f(x) = (x^2 - 2)^2$

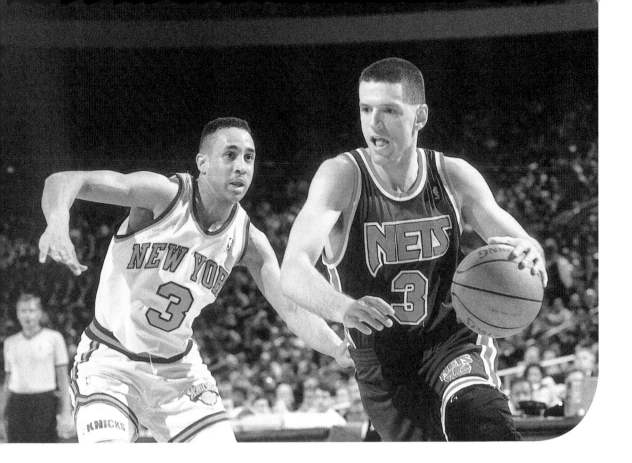

Probability as a tool to make informed decisions

A probability distribution is a mathematical model that shows the possible outcomes of a particular event or course of action as well as the statistical likelihood of each event. For example, a large company might use statistical techniques to create scenario analyses. A scenario analysis uses probability distributions to produce several theoretically distinct possibilities for the outcome of a particular course of action or future event. For example, a business might create three scenarios: worst-case, likely, and best-case. The worst-case scenario would contain a value from the lower end of the probability distribution; the likely scenario would contain a value towards the middle of the distribution; and the best-case scenario would contain a value in the upper end of the distribution.

Although it is impossible to predict the precise value of a future sales level, businesses still need to be able to plan for future events. Using a scenario analysis based on a probability distribution can help a company frame its possible future values in terms of a likely sales level and a worst-case and best-case scenario. By doing so, the company can base its business plans on the likely scenario but still be aware of the alternative possibilities.

? In the last few decades, Probability and Statistics have become very important due to their wide-ranging applications. Statistics literacy is essential not only for business and economics professionals but also, for example, people involved in pro sports. Team coaches often use statistics to decide which players are doing well and then try to predict which players will bring the best results for the game.

> **?** All the probability models that we study in this book are part of the
> Actuarial science syllabus that many universities offer as undergraduate
> or postgraduate studies. Actuarial science combines knowledge from many
> different subjects like mathematics, finance, economics, and computer
> programming. It uses statistical and mathematical methods in insurance,
> health care and finance business models. Working as an actuary has
> become one of the most popular jobs in recent times. In the late 17th
> century there was an advance in research which better established
> probability distributions. Actuarial science immediately used these new
> theories to better formalize fields such as Life insurance and Annuities.
> British mathematician James Dodson (1705–1757), who was a student of
> Abraham de Moivre (1667–1754), was one of the pioneers in Actuarial
> science. Dodson worked as a teacher and an accountant. He modified
> statistical mortality tables developed earlier by Edmund Halley (1656–1742),
> and also formed a society offering the public a more equitable life
> assurance. His plan was put in practice by The Equitable Life, a well-known
> life insurance company established by a group of mathematicians in 1762.

1.1 Cumulative distribution function

You have studied probability distributions as part of the higher level
core course. Before exploring this topic further, it is important that
you revise this part of the core material, particularly if you feel that you
cannot recall the terminology used in this chapter.

Discrete and continuous quantities

The **data** that we collect can be described as **quantitative** or **qualitative**.
Quantitative data can be represented by random variables and classified
into two categories: **discrete** and **continuous**. Discrete random variables
take exact values from the given finite or countable set, whilst the values
of a continuous random variable cannot be listed since they come from an
uncountable set, usually in a form of an interval.

A **discrete** random variable obtained from a finite set of values
$\{x_1, x_2, x_3, ..., x_n\}, n \in \mathbb{Z}^+$ has a *probability distribution function* usually given
by a table of values, listing the values that the variable can take along
with their corresponding probabilities.

x_1	x_2	x_3	x_4	...	x_n
p_1	p_2	p_3	p_4	...	p_n

Sometimes we assign a formula or rule for calculating probabilities.

This is usually when we have an infinite, countable set of values,
$\{x_1, x_2, x_3, ..., x_n, ...\}$, $P(X = x_n) = p_n = p(n)$, where p_n is calculated in
terms of n.

In both cases the following properties must hold:

i $0 \leq p_i \leq 1$

ii $\sum p_i = 1$

A **continuous** random variable obtained from an uncountable set of values, usually written in the form of an interval of real numbers, $[a, b]$, $a < b$, $a, b \in \mathbb{R}$, has a *probability distribution function* given by the formula $f(x) = \begin{cases} f_p(x), \ a \leq x \leq b \\ \quad 0, \text{elsewhere} \end{cases}$

The probability density function must satisfy the following properties:

i $f(x) \geq 0$ for all the values of x

ii $\int\limits_{-\infty}^{+\infty} f(x) = 1$

A **cumulative distribution function** is the sum of all the probabilities of a random variable up to and including the given value of the variable X. In the core course we defined a cumulative distribution function (CDF) of a discrete random variable and used a graphical calculator with CDF features to calculate properties of some discrete probability distributions.

> The word *cumulative* means 'increasing in quantity by successive addition'.

In this chapter we are going to further explore some discrete variables, and study new theoretical discrete probability distributions. We are going to focus on the study of continuous probability distributions.

Even though there is a difference between how we apply the CDF to discrete and to continuous variables, there are also some common properties of the cumulative distribution function that we will look at which are irrespective of the nature of the variable.

> **Definition**
>
> Given the random variable X (discrete or continuous) and the corresponding probability distribution function $P : \mathbb{R} \to [0, 1]$, the *cumulative probability function* $F : \mathbb{R} \to [0, 1]$ is $F(x) = P(X \leq x)$.

This cumulative probability function has the properties:

i $F(x) \in [0, 1]$; i.e. the range of F is $[0, 1]$

ii $\lim\limits_{x \to -\infty} F(x) = 0$

iii $\lim\limits_{x \to +\infty} F(x) = 1$

iv $F(x)$ is nondecreasing on the whole domain; i.e.
$$x_1 < x_2 \Rightarrow F(x_1) \leq F(x_2)$$

Example 1

Prove the cumulative distribution function has the following properties:

a $x_1 < x_2 \Rightarrow P(x_1 < X \leq x_2) = F(x_2) - F(x_1)$,

b $P(X > x) = 1 - F(x)$.

---

a $P(x_1 < X \leq x_2) = P\left(\underbrace{(X \leq x_2)}_{B} \setminus \underbrace{(X \leq x_1)}_{A}\right)$

$\phantom{P(x_1 < X \leq x_2)} = P(X \leq x_2) - P(X \leq x_1)$

$\phantom{P(x_1 < X \leq x_2)} = F(x_2) - F(x_1)$

Set A is a subset of set B so the probability of the difference of these two sets is the difference of the probabilities.
Use the definition of the cumulative distribution function.

b $P(X > x) = \lim_{b \to +\infty} P(x < X \leq b)$

$ = \lim_{b \to +\infty} P(X \leq b) - P(X \leq x)$

$ = \lim_{b \to +\infty} F(b) - F(x) = 1 - F(x)$

Rewrite the given set.

*Use the result from part **a**.*

Use property iii at the bottom of page 6.

In the case of a discrete variable, the cumulative distribution function can be found by simply adding up the values in the table instead of trying to find a generating formula. Just a few discrete variables will have a simple formula; most other formulas are beyond the scope of this book. Therefore we are going to use a table of values for finding the cumulative distribution function for discrete random variables.

Example 2

The probability density function of a random variable X is given by the formula:

$$P(X = x) = \begin{cases} \dfrac{x+k}{9}, & x = 1, 2, 3 \\ 0, & \text{otherwise} \end{cases}$$

a Find the value of k.
b Hence determine the cumulative distribution function.

---

a $\sum_{x=1}^{3} P(X = x) = 1 \Rightarrow$

$\dfrac{1+k}{9} + \dfrac{2+k}{9} + \dfrac{3+k}{9} = 1 \Rightarrow \dfrac{6+3k}{9} = 1$

$k = 1$

The sum of all probabilities must be equal to 1.

Solve the equation and find the value of k.

b $P(X = x) = \dfrac{x+1}{9}, \; x = 1, 2, 3$

$X = x$	1	2	3
$P(X = x)$	$\dfrac{2}{9}$	$\dfrac{3}{9}$	$\dfrac{4}{9}$
$F(x)$	$\dfrac{2}{9}$	$\dfrac{5}{9}$	$\dfrac{9}{9} = 1$

*Use the result from part **a** and calculate the values of the probability density function for x = 1, 2, 3.*

Then use the definition of the cumulative distribution function and add up all the previous probabilities.

Example 3

The cumulative distribution function of a random variable X is given by the table below:

$X = x$	1	2	3	4
$F(x)$	$\dfrac{1}{10}$	$\dfrac{3}{10}$	$\dfrac{3}{5}$	1

Determine the formula of the probability density function.

$P(X = 1) = F(1) = \dfrac{1}{10}$

Use the result from Example 1 to find all the probabilities.

$P(X = 2) = F(2) - F(1) = \dfrac{3}{10} - \dfrac{1}{10} = \dfrac{2}{10}$

$P(X = 3) = F(3) - F(2) = \dfrac{3}{5} - \dfrac{3}{10} = \dfrac{3}{10}$

$P(X = 4) = F(4) - F(3) = 1 - \dfrac{3}{5} = \dfrac{2}{5} = \dfrac{4}{10}$

$f(x) = \begin{cases} \dfrac{x}{10}, & x = 1, 2, 3, 4 \\ 0, & \text{otherwise} \end{cases}$

Look at the repeating pattern and deduce a rule for the values 1, 2, 3, 4.

For continuous random variables the probability density function f and the cumulative distribution function F are related as follows:

$$F'(x) = f(x) \Leftrightarrow F(x) = \int_{-\infty}^{x} f(t)dt$$

The following examples show how to use this relationship and the properties of the cumulative distribution function to solve problems involving continuous random variables.

> Even though the lower boundary of the integral is $-\infty$, in most of the integrals for the left boundary we are going to use the left boundary of the interval of the variable x for which $f(x)$ is not equal to zero.

Example 4

A continuous random variable X has a probability density function given by

$$f(x) = \begin{cases} ax(2-x), & x \in [0, 2] \\ 0, & \text{otherwise} \end{cases}$$

a Find the value of a.

b Hence determine the cumulative distribution function.

c Find the modal value of the random variable X.

---

a $\displaystyle\int_0^2 ax(2-x)\,dx = 1 \Rightarrow a\int_0^2 (2x - x^2)\,dx = 1$

$a\left[x^2 - \dfrac{x^3}{3} \right]_0^2 = 1 \Rightarrow a\left(2^2 - \dfrac{2^3}{3} - 0 \right) = 1$

$\Rightarrow \dfrac{4}{3}a = 1 \Rightarrow a = \dfrac{3}{4}$

The definite integral on the interval [0, 2] must be equal to 1.

Solve the equation and find the value of a.

b $F(x) = \displaystyle\int_0^x f(t)\,dt$

$F(x) = \displaystyle\int_0^x \left(\dfrac{3}{2}t - \dfrac{3}{4}t^2 \right) dt$

$$F(x) = \begin{cases} 0, & x < 0 \\ \dfrac{3}{4}x^2 - \dfrac{x^3}{4}, & 0 \le x \le 2 \\ 1, & x > 2 \end{cases}$$

Use the relationship between probability density and cumulative distribution functions.

*Use the result from part **a** and the properties of the cumulative distribution function and find the formula.*

c $f(x) = \dfrac{3}{2}x - \dfrac{3}{4}x^2 \Rightarrow f'(x) = \dfrac{3}{2} - \dfrac{3}{2}x$

$\dfrac{3}{2} - \dfrac{3}{2}x = 0 \Rightarrow x = 1$

The modal value is the value for which the probability density function reaches an absolute maximum value.

The modal value of the variable X is 1.

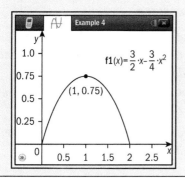

Example 5

A continuous random variable X has a cumulative distribution function, $F(x)$, given by:

$$F(x) = \begin{cases} 0, & x < 0 \\ x^4 - 4x^3 + 4x^2, & 0 \le x \le b \\ 1, & x > b \end{cases}$$

a Find the value of b.
b Hence determine the probability density function.
c Find the median value of the random variable X.

- -

a $b^4 - 4b^3 + 4b^2 = 1$

$\Rightarrow (b^2 - 2b)^2 - 1 = 0$

$\Rightarrow (b^2 - 2b - 1)(b^2 - 2b + 1) = 0$

$b_1 = 1 - \sqrt{2}, \ b_2 = 1 + \sqrt{2}, \ b_{3,4} = 1$

The cumulative function is monotone and there are no points of discontinuity therefore $F(b) = 1$.

Solve the equation and eliminate impossible solutions: b_1 is eliminated because it is negative and b_2 is eliminated because the CDF is not monotone on the interval $[0, 1 + \sqrt{2}]$ as seen on the screen.

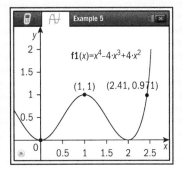

A monotone function is entirely nondecreasing or nonincreasing on the whole domain.

b $f(x) = F'(x)$

$$f(x) = \begin{cases} 4x^3 - 12x^2 + 8x, & x \in [0, 1] \\ 0, & \text{otherwise} \end{cases}$$

*Use the relationship between the probability density and the cumulative distribution functions. Differentiate the probability density function from part **a**.*

c

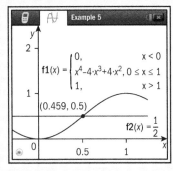

Median, $m = 0.459$.

The graph of the cumulative distribution function of a random variable X is equivalent to the cumulative frequency diagram of a set of data. Notice that both have the same shape (Ogive). To calculate the median value we have to solve the equation $F(m) = \dfrac{1}{2}$, which can be solved on the GDC.

Exercise 1A

1 The probability density function of a random variable X is given by the formula:

$$P(X = x) = \begin{cases} \dfrac{x+k}{12}, & x = 0, 1, 2 \\ 0, & \text{otherwise} \end{cases}$$

a Find the value of k.

b Hence determine the cumulative distribution function.

2 The probability density function of a random variable X is given by the formula:

$$P(X = x) = \begin{cases} \dfrac{a-2x}{20}, & x = 1, 2, 3, 4 \\ 0, & \text{otherwise} \end{cases}$$

a Find the value of a.

b Hence determine the cumulative distribution function.

c Calculate $P(X \le 2)$.

3 The cumulative distribution function of a random variable X is given by the table below:

$X = x$	0	1	2
$F(x)$	$\dfrac{1}{6}$	$\dfrac{1}{2}$	1

a Determine the formula of the probability density function.

b What is the modal value of the variable X?

4 The cumulative distribution function of a random variable X is given by the table below:

$X = x$	1	3	5	7	9
$F(x)$	$\dfrac{1}{25}$	$\dfrac{4}{25}$	$\dfrac{9}{25}$	$\dfrac{16}{25}$	1

a Determine the formula of the probability density function.

b What is the median value of the variable X?

5 A continuous random variable X has a probability density function given by

$$f(x) = \begin{cases} 2\sin bx, & x \in \left[0, \dfrac{\pi}{3}\right] \\ 0, & \text{otherwise} \end{cases}$$

a Find the value of b.

b Hence determine the cumulative distribution function.

c Calculate $P\left(X \ge \dfrac{\pi}{6}\right)$

6 A continuous random variable X has a probability density function given by the formula

$$f(x) = \begin{cases} \dfrac{1}{\pi\sqrt{4-x^2}}, & -2 < x < 2 \\ 0, & \text{otherwise} \end{cases}$$

a Show that this probability function is well defined.

b Find the modal value of the random variable X.

For more practice, see exercise 10C on pages 502–503, and exercise 10K on pages 530–531 of the course companion.

1.2 Other probability distributions

Recall what you have studied in the higher level core course about Bernoulli experiments. A Bernoulli trial has two possible outcomes: success (S) or failure (F). You also learned that a binomial distribution arises from a finite sequence of n such independent experiments with each experiment having an equal probability of success. As you will see in this section the geometric and negative binomial distributions also consist of sequences of such independent experiments with constant probability, but these sequences are of infinite length.

Geometric distribution

There is a very popular board game in Europe called 'Ludo' or 'People, don't get angry'.

 The game is available in many different languages (Ludo, Mensch ärgere dich nicht, Člověče nezlob se, Čovječe ne ljuti se, Лудо, Не се сърди човече...)

Each player chooses to play with the figures of one of the four colours. The game is played by each player moving the figures of his colour by as many fields as is given on a die. To start the game each player must obtain a '6' on a die. Players are given three attempts to start the game. The everlasting question is: 'How difficult is it to start the game?' To answer this question, we must ask: 'What is the probability of being able to start the game within the first three attempts?'

Let's look at the possible sequence of Bernoulli trials. We will denote the outcomes of each single trial by S or F (S denotes 'rolling a 6', and F denotes 'not rolling a 6'). A geometric distribution denotes the infinite sequence of such experiments until we reach the first success, e.g. S, FS, FFS, FFFS, FFFFS, FFFFFS......

Again we need to emphasise that each of these experiments (i.e. each roll of the die) is **mutually independent** and that the probabilities of the possible outcomes are always the same. The probability of a success is denoted by p and the probability of a failure is $q = 1 - p$, $0 \le p \le 1$.

🔍 Jacob Bernoulli (1654–1705), one of eight members of the famous Swiss family of mathematicians in the 17th and 18th centuries. He was the first one to do an extensive piece of work on problems like this.

Definition

A discrete random variable X is said to have a *geometric distribution* and we write $X \sim \text{Geo}(p)$ if $P(X = k) = q^{k-1}p$, where $0 \le p \le 1, q = 1 - p, k = 1, 2, 3, 4...$

This definition satisfies the properties of a probability distribution.

i $p, q \in [0, 1] \Rightarrow q^{k-1}p = P(X = k) \in [0, 1]$ for all $k \in \mathbb{Z}^+$ - every single probability is within the interval $[0, 1]$.

ii $\sum_{k \in \mathbb{Z}^+} P(X = k) = \sum_{k \in \mathbb{Z}^+} q^{k-1}p = p \sum_{k \in \mathbb{Z}^+} q^{k-1} = p\dfrac{1}{1-q} = p\dfrac{1}{p} = 1.$

This means that the sum of all probabilities is 1.

Together, **i** and **ii** show that the above definition for a geometric distribution is a good one.

Example 6

Given that $X \sim \text{Geo}(p)$, find $P(X = k)$ if:

a $p = 0.4, k = 2$
b $p = 0.9, k = 6$
c $p = 0.13, k = 13$

---

a $P(X = 2) = 0.6 \times 0.4 = 0.24$

b $P(X = 6) = 0.1^5 \times 0.9 = 0.000009$

c $P(X = 13) = 0.87^{12} \times 0.13 = 0.0244$

Calculate $q = 1 - p$ and then use the definition of a geometric distribution.

We can calculate those probabilities by using the GDC.

geomPdf(0.4, 2)	0.24
geomPdf(0.9, 6)	0.000009
geomPdf(0.13, 13)	0.024444

3/99

Example 7

Find the probability that a player will successfully start the game "Ludo"
or that the player will obtain a "6" on an unbiased die within the first three attempts.

$X \sim \text{Geo}\left(\dfrac{1}{6}\right)$

Probability of scoring a "6" is $p = \dfrac{1}{6}$.

Method I

$P(X = 1) + P(X = 2) + P(X = 3)$

$= \dfrac{1}{6} + \dfrac{5}{6} \times \dfrac{1}{6} + \left(\dfrac{5}{6}\right)^{2} \times \dfrac{1}{6} = \dfrac{1}{6} + \dfrac{5}{36} + \dfrac{25}{216} = \dfrac{91}{216} = 0.421$

Calculate $q = 1 - p$ and then use the definition of a geometric distribution for values of $k = 1, 2, 3$ and add up all the probabilities.

Method II

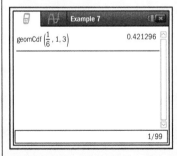

Use a cumulative distribution function on the GDC.

$P(X \leq 3) = 0.421$

Example 8

The probability of starting the game "Ludo" on the third attempt using a biased die is 0.128. Calculate the probability of scoring a "6" on the die.

$X \sim \text{Geo}(p)$ $P(X = 3) = q^2 p$ $(1 - p)^2 p = 0.128$ 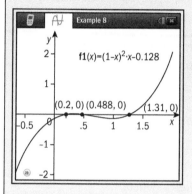 $p_1 = 0.2$, $p_2 = 0.488$, $p_3 = \cancel{1.313}$ Notice that with the given data we have two possible solutions.	*Use the definition of a geometric distribution for value of k = 3 and write the equation in one variable p.* *Use the GDC to solve the equation and eliminate the impossible solution, $0 \le p \le 1$.*

On the GDC we can use two features for solving equations: functions or numerical solver. When solving by the numerical solver we must estimate the values of multiple solutions so that in the iteration process we can input those values to obtain the desirable solution. For the second and third solution we input the values of 0.4 and 1 respectively.

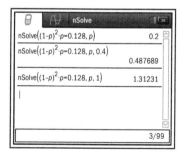

It is much simpler to spot multiple solutions when using the function feature.

Example 9

Find the number of attempts needed such that the chance of successfully starting the game "Ludo" is greater than the probability of not starting the game.

Method I

$$X \sim \text{Geo}\left(\frac{1}{6}\right) \Rightarrow P(X \leq n) > \frac{1}{2}$$

Probability of scoring a "6" within first n attempts should exceed $\frac{1}{2}$. Find the probabilities for values of $k = 1, 2, ..., n$ and add them up.

$$\frac{1}{6} + \frac{5}{6} \times \frac{1}{6} + ... + \left(\frac{5}{6}\right)^{n-1} \times \frac{1}{6} > \frac{1}{2} \Rightarrow$$

$$\frac{1}{6}\left(1 + \frac{5}{6} + ... + \left(\frac{5}{6}\right)^{n-1}\right) > \frac{1}{2} \Rightarrow$$

Notice the geometric sequence and apply the formula for the sum.

$$\frac{1}{6} \times \frac{1 - \left(\frac{5}{6}\right)^{n}}{1 - \frac{5}{6}} > \frac{1}{2} \Rightarrow$$

Simplify the inequality and use logarithms to solve the inequality.

$$\frac{1}{2} > \left(\frac{5}{6}\right)^{n} \Rightarrow \log\left(\frac{1}{2}\right) > n \times \log\left(\frac{5}{6}\right)$$

Logarithms here are negative so don't forget to reverse the inequality symbol.

$$n > \frac{\log\left(\frac{1}{2}\right)}{\log\left(\frac{5}{6}\right)} = 3.80 \Rightarrow n = 4$$

Method II

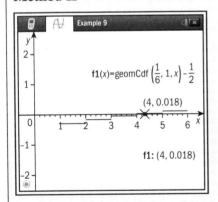

Use the cumulative distribution function on the GDC. Since the values are discrete we obtain a step function and we can find the solution by tracing the graph.

$$P(X \leq n) > 0.5 \Rightarrow n = 4$$

Exercise 1B

1 Given that $X \sim \text{Geo}(p)$ find $P(X = k)$ if:

   a $p = 0.6, k = 2$

   b $p = 0.14, k = 3$

   c $p = 0.5, k = 4$

   d $p = 0.88, k = 5$

2 Given that $X \sim \text{Geo}(p)$ find the following:

   a $P(X \leq 4)$ if $p = 0.25$

   b $P(X > 6)$ if $p = 0.7$

   c $P(5 \leq X \leq 7)$ if $p = 0.3$

   d $P(1 < X \leq 7)$ if $p = 0.991$

3 Mario is shooting at a balloon with a bow and arrow. Each attempt is independent and the probability that Mario hits the balloon with each shot is 0.73. Mario has three arrows. Find the probability that Mario will destroy the balloon.

4 In a certain school 92% of students are familiar with the fire emergency procedure. Students from the school are selected at random. What is the probability that:

   a the fifth selected student is the first one who doesn't know the procedure;

   b the first student selected who doesn't know the procedure will not occur before 4th selected student?

5 Fred manufactures wooden souvenir spoons. The probability that a spoon will be manufactured with no defect is 0.85. Fred inspects every manufactured spoon and when one is found to be defective, he adjusts the machine.

   a What is the probability that the fourth manufactured spoon is the first defective one?

   b What is the probability that within the manufacture of six spoons there will be no need for adjustment?

6 Given that $X \sim \text{Geo}(p)$, show that $P(X > k) = q^k$, $k \in \mathbb{Z}^+$.

Investigation

Let $X \sim \text{Geo}(p)$. Calculate the following probabilities:

a $p = 0.4$, $P(X > 5 \mid X > 3)$, $P(X > 2)$

b $p = 0.7$, $P(X > 6 \mid X > 2)$, $P(X > 4)$

c $p = 0.12$, $P(X > 12 \mid X > 5)$, $P(X > 7)$

Make a conjecture about the connection between the conditional and simple probability, and try to prove the general form of your conjecture.

Expected value and variance of a geometric random variable

In this section we will learn how to find the expected value and the variance of a geometric random variable. In order to do so you will need to manipulate infinite geometric sequences. Recall the formula for the sum of n consecutive terms of a geometric sequence:

$$1 + q + q^2 + q^3 + \dots + q^n = \frac{1 - q^{n+1}}{1 - q}, \text{ where } |q| < 1.$$

When n becomes an extremely large number and q is a number between -1 and 1, the higher powers of q will be very close to zero. Therefore we can write

$$1 + q + q^2 + q^3 + \dots = \lim_{n \to \infty} \left(1 + q + q^2 + q^3 + \dots + q^n\right) = \lim_{n \to \infty} \frac{1 - q^{n+1}}{1 - q} = \frac{1 - 0}{1 - q} = \frac{1}{1 - q}$$

So in other words you have obtained the result for the sum of an infinite geometric sequence.

$$1 + q + q^2 + q^3 + \dots = \frac{1}{1 - q}, \ q \in \left]-1, 1\right[$$

Now if we differentiate this sequence successively, term by term, we will obtain the following:

$$1 + q + q^2 + q^3 + \dots = \frac{1}{1 - q}, \ q \in \left]-1, 1\right[$$

$$\Rightarrow 1 + 2q + 3q^2 + 4q^3 + \dots = \frac{1}{(1 - q)^2}, \ q \in \left]-1, 1\right[$$

If we differentiate again we see that:

$$\Rightarrow 2 + 3 \times 2q + 4 \times 3q^2 + 5 \times 4q^3 + \dots = \frac{2}{(1 - q)^3}, \ q \in \left]-1, 1\right[$$

We will use these results to find the expected value and variance of a geometric random variable, as shown in the following example.

Example 10

Find the expected value and the variance of a geometric random variable X.	
$E(X) = \sum_{k \in \mathbb{Z}^+} k \times P(X = k)$	Use the definition of expected value.
$= 1 \times p + 2 \times qp + 3 \times q^2p + 4 \times q^3p + \ldots + k \times q^{k-1}p + \ldots$	Use the distributive property.
$= p(1 + 2q + 3q^2 + 4q^3 + \ldots + kq^{k-1} + \ldots)$	Use the result obtained in the first derivative of the geometric series and simplify the expression.
$= p\dfrac{1}{(1-q)^2} = p\dfrac{1}{p^2} = \dfrac{1}{p}$	
$\text{Var}(X) = E(X^2) - (E(X))^2 = \sum_{k \in \mathbb{Z}^+} k^2 \times P(X = k) - \left(\dfrac{1}{p}\right)^2$	Use the definition of variance.
$E(X^2) = 1^2 \times p + 2^2 \times qp + 3^2 \times q^2p + 4^2 \times q^3\,p +$ $\ldots + k^2 \times q^{k-1}p + \ldots$	To simplify the process we find $E(X^2)$ first.
$= p(1^2 + 2^2 \times q + 3^2 \times q^2 + 4^2 \times q^3 + \ldots + k^2 \times q^{k-1} + \ldots)$	
$= p((2-1) \times 1 + (3-1) \times 2q + (4-1) \times 3q^2 + (5-1) \times 4q^3 +$ $\ldots + ((k+1) - 1) \times kq^{k-1} + \ldots)$	Use the formula for all the terms. $n^2 = ((n+1) - 1) \times n$
$= p((2 + 3 \times 2q + 4 \times 3q^2 + 5 \times 4q^3 + \ldots + (k+1)k \times q^{k-1} + \ldots)$ $- (1 + 2 \times q + 3 \times q^2 + 4 \times q^3 + \ldots + k \times q^{k-1} + \ldots))$	Rewrite by using the two infinite sums obtained above.
$= p\left(\dfrac{2}{(1-q)^3} - \dfrac{1}{(1-q)^2}\right) = p\left(\dfrac{2}{p^3} - \dfrac{1}{p^2}\right) = \dfrac{2-p}{p^2}$	Use the results of the first and the second derivative to simplify.
$\text{Var}(X) = E(X^2) - (E(X))^2 = \dfrac{2-p}{p^2} - \dfrac{1}{p^2} = \dfrac{1-p}{p^2} = \dfrac{q}{p^2}$	Use the result to find the variance and simplify.

Given a geometric random variable, if $X \sim \text{Geo}(p)$,
then $E(X) = \dfrac{1}{p}$ and $\text{Var}(X) = \dfrac{q}{p^2}$.

This will be used in Example 11 on the next page.

Example 11

Find the expected number of attempts to start the game "Ludo". Use the empirical rule to determine the maximum number of attempts it will take to start the game.

$X \sim \text{Geo}\left(\dfrac{1}{6}\right) \Rightarrow \text{E}(X) = \dfrac{1}{\frac{1}{6}} = 6$	*Apply the formula for the expected value given above.*
$\text{Var}(X) = \dfrac{\frac{5}{6}}{\left(\frac{1}{36}\right)^2} = 30 \Rightarrow \sigma = \sqrt{30} = 5.48$	*Apply the formula for the variance and calculate the standard deviation.*
$[6 - 3 \times 5.48, 6 + 3 \times 5.48] = [-10.4, 22.4]$	*The empirical rule (99.73%) states that the whole population will fall within 3 standard deviations from the mean value, therefore the maximum number of attempts to start the game is 23.*

Exercise 1C

1 Find the expected value and the variance for the geometric distributions given in Exercise 1B, questions **1** and **2**.

2 Mario is shooting at a balloon with a bow and arrow. Each attempt is independent and the probability that Mario hits the balloon with each shot is 0.73.
   a What is the expected number of shots Mario must make to destroy the balloon?
   b Use the empirical rule to find the maximum number of shots Mario must make to destroy the balloon.

3 In a certain school 54% of students are familiar with the election procedure. Students from that school are selected at random. Using the empirical rule, determine how many students must be selected at random to ensure that one of the selected students will be familiar with the election procedure.

Negative binomial distribution

? The negative binomial distribution is also known as Pascal's distribution, named after Blaise Pascal (1623–1662). Pascal was the first mathematician to explore the negative binomial distribution using an integer parameter. The negative binomial distribution is sometimes called Pólya's distribution, named after George Pólya (1887–1985) who extended the parameter values to the set of real numbers.

Let's again consider a sequence of Bernoulli trials but this time we will stop not after the *first* success, but when *two* successes have occurred, e.g. SS, FSS, SFS, FFSS, FSFS, SFFS, FFFSS, FFSFS, FSFFS, SFFFS,.......
In the table below, you can see that for each number of trials, there is now more than one permutation in order to achieve two successes:

Geometric distribution	Two successes
S	SS
FS	FSS, SFS
FFS	FFSS, FSFS, SFFS
FFFS	FFFSS, FFSFS, FSFFS, SFFFS

Notice also that the minimum number of trials we must perform is the number of successes we require, which in this case is two.

Let's consider a sequence of Bernoulli trials where we stop when three successes have occurred. To make it simpler, we'll use the table below to compare the number of different permutations available for three successes within the geometric distribution.

Geometric distribution	Three successes
S	SSS
FS	FSSS, SFSS, SSFS,
FFS	FFSSS, FSFSS, FSSFS, SFFSS, SFSFS, SSFFS

We might notice that as we increase the number of successes required, we increase the number of permutations available.

In general, if we record up to r successes in a sequence of k Bernoulli trials we actually consider two independent events. We need to obtain $r-1$ successes in the first $k-1$ experiments **and** we need to obtain one last success in the k-th trial. The first part of the sequence can be described by the binomial distribution with the parameters $k-1$ and $r-1$, and then at the end we need to have one more success. So we can find the probability density function.

$$P(X = k) = \underbrace{\binom{k-1}{r-1}q^{k-r}p^{r-1}}_{r-1 \text{ successes within } k-1 \text{ trials}} \times \underbrace{p}_{k\text{-th trial } r\text{-th success}} = \binom{k-1}{r-1}q^{k-r}p^{r}$$

In order to have r successes we must have at least r experiments, therefore $k = r, r+1, r+2, r+3, \ldots$

This discrete distribution is called negative binomial distribution.

Definition

A discrete random variable X is said to have *negative binomial distribution* and we write $X \sim \text{NB}(r, p)$ if $P(X = k) = \binom{k-1}{r-1}q^{k-r}p^{r}$, where $0 \leq p \leq 1$, $q = 1 - p$, $k = r, r+1, r+2, r+3, \ldots$

It is left as an exercise to show that this is a good definition of a probability distribution. Therefore we need to prove that:

i $P(X = k) \in [0, 1] \Rightarrow 0 \le \binom{k-1}{r-1} q^{k-r} p^r \le 1$

for all $k = r, r+1, r+2, r+3, \ldots$

ii $\sum_{k=r}^{\infty} P(X = k) = 1 \Rightarrow \sum_{k=r}^{\infty} \binom{k-1}{r-1} q^{k-r} p^r = 1$

> **?** Geometric distribution is in fact a special case of a negative binomial distribution where we require just one success, $r = 1$, $\text{Geo}(p) = \text{NB}(1, p)$.

Example 12

Given that $X \sim \text{NB}(r, p)$ find $P(X = k)$ if:

a $r = 2, p = 0.1, k = 2$
b $r = 6, p = 0.5, k = 9$
c $r = 12, p = 0.25, k = 50$

---

a $P(X = 2) = \binom{2-1}{2-1} 0.9^{2-2} \times 0.1^2 = 0.01$

b $P(X = 9) = \binom{9-1}{6-1} 0.5^{9-6} \times 0.5^6 = 0.109$

c $P(X = 50) = \binom{50-1}{12-1} 0.75^{50-12} \times 0.25^{12} = 0.0310$

Use the definition of a negative binomial distribution and apply the formula.

Unfortunately, graphic display calculators have no negative binomial distribution under the distribution features, but the programming capacity of a calculator helps us to write down a useful program with just a few lines.

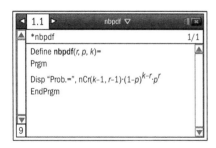

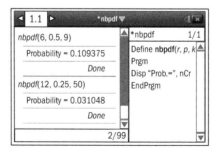

A similar programme can be created for the cumulative distribution function of a negative binomially distributed variable.

Example 13

> A new drug is to be tested on people. If only 12% of the people asked to participate agree to take part in the study, what is the probability that:
> **a** 8 people will be asked before 3 are found who agree to participate;
> **b** 12 people will be asked before 5 are found who agree to participate;
> **c** not more than 5 people are asked before 2 are found who agree to participate?

$X \sim \mathrm{NB}(r, p)$

a $r = 3$, $p = 0.12 \Rightarrow P(X = 8) = \binom{7}{2} 0.88^{5} 0.12^{3}$

$\quad = 0.0192$

Identify the parameters of a negative binomial distribution and use the formula to calculate the probabilities.

b $r = 5$, $p = 0.12 \Rightarrow P(X = 12) = \binom{11}{4} 0.88^{7} 0.12^{5}$

$\quad = 0.00336$

c $r = 2$, $p = 0.12 \Rightarrow P(2 \le X \le 5)$

$\quad = P(X = 2) + P(X = 3) + P(X = 4) + P(X = 5)$

$\quad = \binom{1}{1} 0.12^{2} + \binom{2}{1} 0.88 \times 0.12^{2} + \binom{3}{1} 0.88^{2} \times 0.12^{2} +$

$\quad \binom{4}{1} 0.88^{3} \times 0.12^{2}$

$\quad = 0.112$

Identify the values of the variable and add up all the probabilities for the identified values.

> **?** The binomial and negative binomial distributions are similar: the binomial distribution takes a fixed number n of Bernoulli trials and generates probabilities of the number of successes, r, as r varies from 0 to n. Conversely, negative binomial distribution takes a fixed number of successes, r, and generates probabilities that we perform n Bernoulli trials (where n varies from r to infinity) in order to achieve these r successes.

Exercise 1D

1 Given that $X \sim \mathrm{NB}(r, p)$ find $P(X = k)$ if:
   a $r = 1$, $p = 0.2$, $k = 2$
   b $r = 3$, $p = 0.5$, $k = 4$
   c $r = 7$, $p = 0.8$, $k = 9$
   d $r = 23$, $p = 0.77$, $k = 32$

2 Given that $X \sim \mathrm{NB}(r, p)$ find:
   a $P(X \le 4)$ if $p = 0.25$ and $r = 3$
   b $P(X > 6)$ if $p = 0.5$ and $r = 2$
   c $P(5 \le X \le 7)$ if $p = 0.82$ and $r = 4$
   d $P(8 < X \le 11)$ if $p = 0.43$ and $r = 5$

3 The random variable has the following distribution $X \sim \text{NB}(2, p)$.

    a Given that $p < \dfrac{1}{2}$ and $P(X = 3) = \dfrac{24}{125}$, find the value of p.

    b Hence find $P(3 \le X \le 5)$.

4 Before Alemka can start a game, she has to obtain two "oncs" when rolling a fair tetrahedral die with the standard faces 1, 2, 3, and 4. Let random variable X denote the total number of times Alemka has to throw the die until she obtains the second "one".

    a Write the distribution of X, including the value(s) of each parameter.

    b Find the value of x such that $P(X = x) = \dfrac{3}{32}$.

    c Calculate $P(X \le 5)$.

5 Nicholas is playing a game on his smart phone. In this game he needs to eat apples in order to pass to the next level. The probability that Nicholas eats an apple is 0.85. He needs to eat at least four apples in order to advance in the game. What is the probability that he will eat

    a exactly five apples before he advances;

    b at least seven apples before he advances?

6 In a certain school 92% of students are familiar with the fire emergency procedure. Students from that school are selected at random. The fire inspector conducts an interview and needs at least five students who are familiar with the procedure. What is the probability that:

    a he will need to interview exactly six students in order to satisfy the need;

    b he will not need to interview more than a dozen students?

7 An ice-cream factory produces a special type of a diet ice-cream. The probability that an ice-cream will be produced with no defect in shape is 0.85. Tony inspects every ice-cream and when he finds three ice-creams to have a shape defect he adjusts the machine.

    a What is the probability that he will inspect exactly five ice-creams before he adjusts the machine?

    b What is the probability that within half a dozen ice-creams there will be no need for adjustment?

1.3 Probability generating functions

When tossing two coins the number of heads obtained is recorded. Let X be the discrete random variable. The table below shows the probability distribution function of X.

$X = x_i$	0	1	2
$P(X = x_i)$	$\frac{1}{4}$	$\frac{1}{2}$	$\frac{1}{4}$

You may notice that, in this table, the probabilities are arranged sequentially as the value of X changes. From this table we can write the probabilities in the form of a polynomial expression involving a variable, t:

$$G(t) = \frac{1}{4} \times t^0 + \frac{1}{2} \times t^1 + \frac{1}{4} \times t^2 = \frac{1}{4} + \frac{1}{2}t + \frac{1}{4}t^2$$

This expression is called a **probability generating function** for this discrete random variable. We can immediately notice that $G(1) = \frac{1}{4} + \frac{1}{2} + \frac{1}{4} = 1$ since the coefficients of this polynomial are the corresponding probabilities and the sum of all the probabilities must be equal to 1.

When studying Probability it is important not to rely too much on intuition. Even the sharpest minds like Jean D'Alembert (1717–1783) have at some stage made famous mistakes. In his article Croix ou Pile in the French Encyclopédie he considered the following problem: in two tosses of a fair coin, what is the probability that Heads will appear at least once? D'Alembert's answer was $\frac{2}{3}$. He reasoned that in real life no one would continue the experiment after Heads showed up on the first toss. In other words, D'Alembert mistakenly assumed that the sample space was {H,TH,TT}.

Example 14

A pair of dice are rolled. Let X denote the sum of the outcomes on the upper faces. Write the probability distribution function of the random variable X and find the probability generating function.

x_i	2	3	4	5	6	7	8	9	10	11	12
p_i	$\frac{1}{36}$	$\frac{2}{36}$	$\frac{3}{36}$	$\frac{4}{36}$	$\frac{5}{36}$	$\frac{6}{36}$	$\frac{5}{36}$	$\frac{4}{36}$	$\frac{3}{36}$	$\frac{2}{36}$	$\frac{1}{36}$

$$G(t) = \frac{1}{36}t^2 + \frac{1}{18}t^3 + \frac{1}{12}t^4 + \frac{1}{9}t^5 + \frac{5}{36}t^6 + \frac{1}{6}t^7 + \frac{5}{36}t^8$$

$$+ \frac{1}{9}t^9 + \frac{1}{12}t^{10} + \frac{1}{18}t^{11} + \frac{1}{36}t^{12}$$

Look at all the possible pairs of outcomes and their sums. Find the corresponding probabilities. Take p_i to be the coefficients of the polynomial and x_i to be the powers.

Again notice that $G(1) = 1$.

A random variable can also assume values in an infinite set of numbers as in the case of a geometric random variable, as shown in Example 15.

Example 15

A coin is flipped until a head is obtained. The random variable X denotes the number of flips. Write the probability distribution of the random variable describing this experiment and find the probability generating function.

x_i	1	2	3	4	...	k	...
p_i	$\dfrac{1}{2}$	$\dfrac{1}{4}$	$\dfrac{1}{8}$	$\dfrac{1}{16}$	...	$\dfrac{1}{2^k}$	...

The possible outcomes are H, TH, TTH, TTTH,... and so on. $X \sim Geo\left(\dfrac{1}{2}\right)$.

$$G(t) = \frac{1}{2}t + \frac{1}{4}t^2 + \frac{1}{8}t^3 + \frac{1}{16}t^4 + ... + \frac{1}{2^k}t^k + ...$$

$G(t)$ is an infinite geometric series with common ratio $\dfrac{1}{2}t$.

$$= \frac{\frac{1}{2}t}{1 - \frac{1}{2}t} = \frac{t}{2-t}, \ t \neq 2$$

Use the formula for the sum of an infinite geometric series and simplify the expression.

Notice that we were using the formula $S_\infty = \dfrac{u_1}{1-r}$ for the sum of an infinite geometric series, under the condition that this series converges.

That is, we must have $-1 < \dfrac{1}{2}t < 1 \Rightarrow -2 < t < 2$. For the other values of t, this sum is not defined.

Again we also notice that $G(1) = \dfrac{1}{2-1} = 1$.

Definition

Let X be a discrete random variable assuming nonnegative integer values and $P(X = k) = p_k$, $k = 0, 1, 2, 3...$ are the corresponding probabilities. Then a function $G : R \rightarrow R$ of the form $G(t) = \displaystyle\sum_{k=0}^{\infty} p_k t^k = p_0 + p_1 t + p_2 t^2 + p_3 t^3 + ... + p_n t^n + ...$ is called a *probability generating function*.

As seen in the previous examples, if a discrete random variable takes values only within a finite set, we could consider this as an infinite set where the probability of the variable taking any value outside the finite set is zero.

Let's try to find probability generating functions for some discrete distributions that we are already familiar with.

Bernoulli distribution $X \sim B(1, p)$:

$(P(X = 0) = q)$ and $(P(X = 1) = p) \Rightarrow G(t) = qt^0 + pt^1 = q + pt$

Binomial distribution $X \sim B(n, p)$:

$$P(X = k) = \binom{n}{k} q^{n-k} p^k, \ k = 0, 1, 2, 3, ..., n \Rightarrow G(t)$$

$$= \sum_{k=0}^{n} \binom{n}{k} q^{n-k} p^k t^k = \sum_{k=1}^{n} \binom{n}{k} q^{n-k} (pt)^k = (q + pt)^n$$

Poisson distribution $X \sim \text{Po}(m)$:

$$P(X = k) = \frac{m^k e^{-m}}{k!}, k = 0, 1, 2, 3, \ldots \Rightarrow G(t) = \sum_{k=0}^{\infty} \frac{m^k e^{-m}}{k!} t^k$$

$$= e^{-m} \sum_{k=0}^{\infty} \frac{(mt)^k {}^*}{k!} = e^{-m} \times e^{mt} = e^{m(t-1)}$$

> *We have used Maclaurin's formula for the exponential function $e^x = \sum_{k=0}^{\infty} \frac{x^k}{k!}$ This formula is a part of the Calculus option.

Geometric distribution $X \sim \text{Geo}(p)$:

$$P(X = k) = q^{k-1} p, k = 1, 2, 3, \ldots \Rightarrow G(t) = \sum_{k=1}^{\infty} q^{k-1} p t^k = pt \sum_{k=1}^{\infty} q^{k-1} t^{k-1}$$

Substitute $k - 1 = i$ then,

$$pt \sum_{k=1}^{\infty} q^{k-1} t^{k-1} = pt \sum_{i=0}^{\infty} q^i t^i = \frac{pt}{1 - qt}$$

When $-1 < qt < 1$, this is an infinite geometric series with the common ratio qt.

Negative binomial distribution $X \sim \text{NB}(r, p)$:

$$P(X = k) = \binom{k-1}{r-1} q^{k-r} p^r, k = r, r+1, r+2, \ldots \Rightarrow G(t) = \sum_{k=r}^{\infty} \binom{k-1}{r-1} q^{k-r} p^r t^k$$

$$G(t) = p^r t^r \sum_{k=r}^{\infty} \binom{k-1}{r-1} q^{k-r} t^{k-r} = p^r t^r \sum_{i=0}^{\infty} \binom{r+i-1}{i} q^i t^i$$

Using the substitution $k - r = i$ and the formula for negative

binomial series $(1 - x)^{-n} = \sum_{i=0}^{\infty} \binom{n+i-1}{i} x^i$, where $-1 < qt < 1$.

$$p^r t^r \sum_{i=0}^{\infty} \binom{r+i-1}{i} q^i t^i = \frac{p^r t^r}{(1 - qt)^r} = \left(\frac{pt}{1 - qt} \right)^r$$

Example 16

A random variable X has a probability generating function $G(t) = \dfrac{a}{4 - t}, t \neq 4$.
a Find the value of a.
b Hence calculate $P(1 \leq X \leq 3)$.

a $\dfrac{a}{4 - 1} = 1 \Rightarrow a = 3$	*Use the fact that $G(1) = 1$.*
b $G(t) = \dfrac{3}{4 - t} = \dfrac{3}{4} \times \dfrac{1}{1 - \dfrac{t}{4}}$	*Rewrite it in a polynomial form.*
$= \dfrac{3}{4} \times \left(1 + \dfrac{1}{4} t + \dfrac{1}{16} t^2 + \dfrac{1}{64} t^3 + \ldots \right)$	
$P(1 \leq X \leq 3) = \dfrac{3}{4} \times \left(\dfrac{1}{4} + \dfrac{1}{16} + \dfrac{1}{64} \right) = \dfrac{63}{256}$	*Add the coefficients of the powers 1, 2, and 3.*

Now let's take a look at the probability generating function and its successive derivatives.

$$G(t) = \sum_{k=0}^{\infty} p_k t^k = p_0 + p_1 t + p_2 t^2 + p_3 t^3 + \ldots + p_n t^n + \ldots$$

$$G'(t) = 0 + p_1 + 2p_2 t + 3p_3 t^2 + 4p_4 t^3 + \ldots + np_n t^{n-1} + \ldots = \sum_{k=1}^{\infty} k \times p_k t^{k-1}$$

$$G''(t) = 0 + 0 + 2p_2 + 3 \times 2p_3 t + 4 \times 3p_4 t^2 + \ldots + n(n-1)p_n t^{n-2} + \ldots = \sum_{k=2}^{\infty} k(k-1) \times p_k t^{k-2}$$

By considering the derivatives of the probability generating function when $t = 1$, we can derive some very useful results for the value of $t = 1$:

i $\quad G(1) = \sum_{k=0}^{\infty} p_k 1^k = \sum_{k=0}^{\infty} p_k = 1$

ii $\quad G'(1) = \sum_{k=1}^{\infty} k p_k 1^{k-1} = \sum_{k=1}^{\infty} k p_k = \mathrm{E}(X)$

iii $\quad G''(1) = \sum_{k=2}^{\infty} k(k-1) p_k 1^{k-2} = \sum_{k=2}^{\infty} k(k-1) p_k = \mathrm{E}(X(X-1))$

As you can see, we can calculate the expected value $\mathrm{E}(X)$ directly from the first derivative of a probability generating function. We can also calculate the variance $\mathrm{Var}(X)$ of a random variable using derivatives of the probability generating function.

$$G''(1) = \sum_{k=2}^{\infty} k(k-1) p_k = \sum_{k=2}^{\infty} (k^2 - k) p_k = \sum_{k=2}^{\infty} k^2 p_k - \sum_{k=2}^{\infty} k p_k + p_1 - p_1 = \sum_{k=1}^{\infty} k^2 p_k - \sum_{k=1}^{\infty} k p_k$$

$$= \mathrm{E}(X^2) - \mathrm{E}(X)$$

$$G''(1) = \mathrm{E}(X^2) - \underbrace{\mathrm{E}(X)}_{G'(1)} \Rightarrow \mathrm{E}(X^2) = G''(1) + G'(1)$$

$$\mathrm{Var}(X) = \mathrm{E}(X^2) - (\mathrm{E}(X))^2 = G''(1) + G'(1) - (G'(1))^2 = G''(1) + G'(1)(1 - G'(1))$$

Let's summarize the results we just obtained:

$$\begin{array}{|l|}
\hline
G(1) = 1 \\
\mathrm{E}(X) = G'(1) \\
\mathrm{Var}(X) = G''(1) + G'(1)(1 - G'(1)) \\
\hline
\end{array}$$

Knowing these formulas and generating functions we can much more easily calculate some of the expected values and variances that previously were difficult to find. We are going to start with the binomial distribution.

Example 17

Use the results of the probability generating function to find the expected value and the variance of a Binomial random variable with parameters n and p.

$X \sim B(n, p) \Rightarrow G(t) = (q + pt)^n$ $\Rightarrow G'(t) = n(q + pt)^{n-1} \times p = np(q + pt)^{n-1}$ $\Rightarrow G''(t) = np(n - 1)(q + pt)^{n-2} \times p$ $\qquad = n(n - 1)p^2(q + pt)^{n-2}$	Use the probability generating function formula and differentiate it twice to obtain the first and second derivatives.
$E(X) = G'(1) = np\left(\underbrace{q + p \times 1}_{1}\right)^{n-1} = np$	Use the formula for expected value.
$Var(X) = G''(1) + G'(1)(1 - G'(1))$ $\qquad = n(n - 1)p^2\left(\underbrace{q + p \times 1}_{1}\right)^{n-2} + np(1 - np)$	Use the formula for variance and the previous results.
$\qquad = n^2p^2 - np^2 + np - n^2p^2 = np(1 - p) = npq$	Simplify the result.

It is much easier to find the expected value and the variance of a Poisson distribution, as Example 18 shows.

Example 18

Use the probability generating function to find the expected value and the variance of a Poisson random variable with parameter m.

$X \sim Po(m) \Rightarrow G(t) = e^{m(t-1)}$ $\Rightarrow G'(t) = e^{m(t-1)} \times m = me^{m(t-1)}$ $\Rightarrow G''(t) = me^{m(t-1)} \times m = m^2e^{m(t-1)}$	Use the probability generating function formula and differentiate it twice to obtain the first and second derivatives.
$E(X) = G'(1) = me^{m(1-1)} = m$ $Var(X) = G''(1) + G'(1)(1 - G'(1))$ $\qquad = m^2e^{m(1-1)} + me^{m(1-1)}(1 - me^{m(1-1)})$	Use the formula for expected value. Use the formula for variance and the previous results.
$\qquad = m^2 + m - m^2 = m$	Simplify the result.

There are still many other distributions that do not have special names and for many of those distributions, probability generating functions make their calculations much easier. Example 19 will illustrate this.

Example 19

Max and Marco alternately throw darts. The first one who scores a bull's-eye wins the game. The probabilities that Max and Marco score a bull's-eye in every shot are $\frac{3}{4}$ and $\frac{5}{8}$ respectively. Max throws first. Their successive throws are independent and the random variable X denotes the number of throws before the game is over.
a Find the probability generating function of the variable X and verify your result.
b What is the expected number of throws before the game is over?

---

a $G(t) = \frac{3}{4}t + \frac{1}{4} \times \frac{5}{8}t^2 + \frac{1}{4} \times \frac{3}{8} \times \frac{3}{4}t^3 + \frac{1}{4} \times \frac{3}{8} \times \frac{1}{4} \times \frac{5}{8}t^4$

Max scores or
Max misses and Marco scores or
Max misses and Marco misses
and Max scores or
Max misses and Marco misses
and Max misses and Marco
scores or...

$+ \frac{1}{4} \times \frac{3}{8} \times \frac{1}{4} \times \frac{3}{8} \times \frac{3}{4}t^5 + \frac{1}{4} \times \frac{3}{8} \times \frac{1}{4} \times \frac{3}{8} \times \frac{1}{4} \times \frac{5}{8}t^6 + \ldots$

Notice that there are two infinite geometric series that alternate term by term with different first terms and equal common ratio.

$= \frac{3}{4}t \left(1 + \frac{3}{32}t^2 + \left(\frac{3}{32}t^2\right)^2 + \ldots \right)$

$+ \frac{5}{32}t^2 \left(1 + \frac{3}{32}t^2 + \left(\frac{3}{32}t^2\right)^2 + \ldots \right)$

$= \left(\frac{3}{4}t + \frac{5}{32}t^2 \right) \frac{1}{1 - \frac{3}{32}t^2} = \frac{24t + 5t^2}{32} \times \frac{32}{32 - 3t^2} = \frac{24t + 5t^2}{32 - 3t^2}$

Use the formula for the sum of an infinite sequence and simplify your answer.

$G(1) = \frac{24 + 5}{32 - 3} = \frac{29}{29} = 1$

The sum of all probabilities must be equal to 1, G(1) = 1.

b **Method I**

$G(t) = \frac{24t + 5t^2}{32 - 3t^2}$

Differentiate the probability generating function. Use the quotient rule.

$\Rightarrow G'(t) = \frac{(24 + 10t)(32 - 3t^2) - (24t + 5t^2)(-6t)}{(32 - 3t^2)^2}$

$\Rightarrow G'(t) = \frac{768 + 320t - 72t^2 - 30t^3 + 144t^2 + 30t^3}{(32 - 3t^2)^2}$

$= \frac{768 + 320t + 72t^2}{(32 - 3t^2)^2}$

Simplify the result.

$\Rightarrow E(X) = G'(1) = \frac{768 + 320 + 72}{(32 - 3)^2} = \frac{1160}{841} = 1.38$

Calculate G'(1). Notice that we expect at least two throws to be made.

At least two throws are made before the game is over.

Method II

1.1 ▸ *Example 19 $\dfrac{d}{dx}\left(\dfrac{24\cdot x+5\cdot x^2}{32-3\cdot x^2}\right)\Big\|_{x=1}$ $\dfrac{40}{29}$ $\dfrac{40}{29}$ 1.37931 2/99	*Use a GDC to find $G'(1)$.*
At least two throws are made before the game is over.	*Notice that we expect at least two throws to be made.*

Exercise 1E

1   Four coins are tossed. Let X denote the number of heads that appear face up. Write the probability distribution of the random variable X and find the probability generating function.

2   A die is rolled until we get a "1". The random variable X denotes the number of rolls. Write the probability distribution of the random variable describing this experiment and show that the probability generating function is $G(t) = \dfrac{t}{6-5t}$, $t \neq \dfrac{6}{5}$.

3   A random variable X has a probability generating function $G(t) = \dfrac{2}{3-t}$, $t \neq 3$. Find:

   a $P(X = 0)$

   b $P(X \leq 1)$

   c $P(X \geq 3)$

   d $P(X \geq k)$

4   Use the probability generating function to find the expected value and the variance of the following variables:

   a X is a Bernoulli variable with the probability p;

   b X is a negative binomial with the parameters p and r.

5   Find the expected value and the variance of the random variables in questions 1, 2 and 3.

6 Noddy and Eve play a basketball game in their courtyard. Taking turns, they shoot a basketball at a hoop from a free throw line. The first one who makes a basket wins the game. The probabilities that Noddy and Eve score in any shot are $\frac{2}{3}$ and $\frac{4}{7}$ respectively. Successive shots are independent. Eve shoots first. The random variable X denotes the number of shots before the game is over.

  a Find the value of $P(X = 1)$ and show that $P(X = 4) = \frac{2}{49}$.

  b Find the probability generating function for the random variable X and verify your result.

  c What is the expected number of shots before the game is over?

  d Use the empirical rule to find the maximum number of shots to be made before the game is over.

The sum of independent variables

If we look at the experiment of tossing two coins we can say that we actually have two independent events of tossing one coin at a time. Now we are going to note the number of heads and we can say that coin 1 is represented by the random variable X, and coin 2 is represented by the random variable Y. Both X and Y have the same probability distribution:

x_i	0	1
p_i	$\frac{1}{2}$	$\frac{1}{2}$

They also have the same generating function:

$$G_X(t) = G_Y(t) = \frac{1}{2}t^0 + \frac{1}{2}t^1 = \frac{1}{2} + \frac{1}{2}t.$$

We notice that if we multiply these two probability generating functions we obtain the result we calculated earlier, when we found a probability generating function for the number of heads shown when we tossed two coins together.

$$G_X(t) \times G_Y(t) = \left(\frac{1}{2} + \frac{1}{2}t\right) \times \left(\frac{1}{2} + \frac{1}{2}t\right) = \frac{1}{4} + \frac{1}{2}t + \frac{1}{4}t^2 = G_{X+Y}(t)$$

In the course companion, we were easily manipulating the parameter of a Poisson distribution. For example, suppose that the mean number of cars arriving at a petrol station in one hour is m. We were able to say that within half an hour the mean number was $\frac{m}{2}$, and in three hours the mean number was $3m$. The reason why we were able do this is that the parameter of a Poisson distribution is the expected value or the mean. So parameters of the corresponding Poisson distributions were calculated by using the same elementary transformations. We were then finding the given probabilities.

Example 20

The times at which a fly and a wasp arrive into a room can be modelled by Poisson distributions with parameters 0.3 and 0.1 respectively. Insects arrive independently into the room. Let random variables X and Y denote the number of flies and wasps in the room respectively. Let the random variable Z denote the total number of flies and wasps in the room. Find probability generating functions of X, Y and Z, and find the possible relationship between them.

$Z = X + Y$ $$G_X(t) = \sum_r P(X = r)t^r = e^{0.3(t-1)}$$ $$G_Y(t) = \sum_s P(Y = s)t^s = e^{0.1(t-1)}$$	*The number of insects Z is the number of flies X and wasps Y in the room. Write down probability generating functions for each insect. To follow the calculation for Y (the number of wasps) more easily, use a different index, s.*
$$G_Z(t) = \sum_k P(Z = k)t^k = e^{0.4(t-1)}$$ $$= e^{(0.3+0.1)(t-1)} = e^{0.3(t-1)+0.1(t-1)} = e^{0.3(t-1)} \times e^{0.1(t-1)}$$ $$= G_X(t) \times G_Y(t)$$	*When we count the number of both insects in the room we make no distinction between flies and wasps. Therefore notice that $Z \sim Po(0.4)$.*

The example above illustrates a very important property of probability distribution functions that doesn't depend on the probability distribution of the given random variables, but is valid across all probability distributions.

Theorem 1

X_1 and X_2 are two independent random variables with the corresponding probability generating functions $G_{X_1}(t)$ and $G_{X_2}(t)$. If a new random variable X is such that

$X = X_1 + X_2$ then $G_X(t) = G_{X_1 + X_2}(t) = G_{X_1}(t) \times G_{X_2}(t)$.

Proof:

$$G_X(t) = \sum_k P(X = k)t^k$$	
$$= \sum_{r+s} P(X = r + s)t^{r+s}$$	*Use the substitution $k = r + s$.*
$$= \sum_r \sum_s P((X_1 = r) \cap (X_2 = s))t^{r+s}$$	*Rewrite the sum by using both indices and variables X_1 and X_2.*
$$= \sum_r \sum_s P(X_1 = r)P(X_2 = s)t^r t^s$$	*Since the variables are independent use the multiplicative probability law.*
$$= \sum_r P(X_1 = r)t^r \sum_s P(X_2 = s)t^s$$	*Use the distributive property.*
$$= G_{X_1}(t) \times G_{X_2}(t)$$	*Q.E.D.*

Theorem 1 can be extended to find the probability generating function of the sum of more than two independent random variables.

Corollary

Given that $X_1, X_1, ..., X_n, n \in \mathbb{Z}^+$, are n independent random variables with the corresponding probability generating functions $G_{X_1}(t), G_{X_2}(t), ..., G_{X_n}(t)$.

If a new random variable X is such that $X = X_1 + X_2 + ... + X_n = \sum_{k=1}^{n} X_k$ then

$$G_X(t) = G_{X_1}(t) \times G_{X_2}(t) \times ... \times G_{X_n}(t) = \prod_{k=1}^{n} G_{X_k}(t).$$

The proof of this corollary is left to the reader as an exercise.

This corollary helps us to easily calculate the probability generating functions of a random variable which can be modelled as a sequence of simpler experiments described by the same random variable.

Example 21

Random variable X denotes a Bernoulli trial and its corresponding probability generating function is given by $G_X(t) = q + pt$, where p and q are respective probabilities of a success and a failure in the trial. Use the corollary above to find the probability distribution function of the binomial random variable Y that describes a sequence of n such independent trials.

$$Y = \underbrace{X + X + ... + X}_{n \text{ independant trials}} = \sum_{k=1}^{n} X$$

$$\Rightarrow G_Y(t) = \underbrace{G_X(t) \times G_X(t) \times ... \times G_X(t)}_{n \text{ factors}}$$

$$= \underbrace{(q + pt) \times (q + pt) \times ... \times (q + pt)}_{n \text{ factors}}$$

$$= (q + pt)^n$$

A binomial distribution describes a sequence of n independent repetitions of a Bernoulli trial. Therefore Y follows a binomial distribution.

Use the corollary to find the probability generating function.

Notice that we have obtained the same result much simpler.

Exercise 1F

1 The probability generating functions of independent random variables X and Y are given by the following formulas: $G_X(t) = \left(\dfrac{1 + 3t}{4}\right)^2$ and $G_Y(t) = \left(\dfrac{2 + t}{3}\right)^2$.

   a Determine $G_{X+Y}(t)$

   b Find $P(X + Y \leq 1)$

   c Calculate $E(X + Y)$

2 The probability generating functions of independent random variables

X and Y are given by the following formulas:

$$G_X(t) = \left(\frac{t}{2-t}\right)^3 \text{ and } G_Y(t) = \left(\frac{t}{3-2t}\right)^3.$$

a Determine $G_{X+Y}(t)$

b Calculate $E(X+Y)$

c Calculate $\text{Var}(X+Y)$

3 The times in which bikes, cars, and buses arrive at a crossing can be modelled by Poisson distributions with parameters 0.25, 0.15, and 0.05 respectively. All of them arrive independently at the crossing. Let random variables X, Y and Z denote the numbers of each of them at the crossing.

a Find the probability generating functions of X, Y and Z.

b Find the probability that at least two kinds of vehicle will be at the crossing at a given moment.

? Traffic Analysis is a vital component in understanding the requirements and capabilities of a network. There are many traffic models proposed for analyzing the traffic characteristics of networks, but none of them can efficiently capture the traffic characteristics of all types of networks under every possible circumstance. However, one of the most widely used and oldest traffic models is the Poisson Model.

4 Geometric random variable X has the corresponding probability distribution function given by $G_X(t) = \dfrac{pt}{1-qt}$, where p and q are probabilities of a success and a failure respectively. Use the corollary to find the probability distribution function of the negative binomial random variable Y that describes a sequence of independent trials until we achieve r successes.

5 a X_1 and X_2 are independent Binomial variables with the same probability p, however they have different numbers of repetitions, n_1 and n_2 respectively. Use the probability generating function to show that $X_1 + X_2$ is a Binomial variable and hence find the parameters.

b Given that $X_1, X_2, ..., X_k, k \in \mathbb{Z}^+$ are independent Binomial random variables having the same probability p but different number of repetitions, $n_1, n_2, ..., n_k$ respectively. Use mathematical induction to show that $Y = \sum_{i=1}^{k} X_i$ is also a Binomial distribution and find the parameters.

Review exercise

1 A random variable X has a probability generating function
$$G(t) = \frac{4t}{5-t}, t \neq 5. \text{ Find:}$$

   a $P(1 \leq X \leq 4)$

   b $P(X \geq 2)$

   c $E(X)$

   d $Var(X)$

2 The times a rabbit, a fox, and a deer arrive at a meadow can be modelled by Poisson distributions with parameters 0.6, 0.12, and 0.28 respectively. All of them arrive independently at the meadow. Let random variables X, Y, and Z denote the numbers of each of them at the meadow.

   a Find the probability generating functions of X, Y, and Z.

   b Find the expected number of animals at the meadow.

   c Find the probability that at least two animals will be at the meadow at a given moment.

3 In a certain community, 75% of volunteers are involved with the Peer Helping programme. Students from the school are selected at random. What is the probability that:

   a i the fourth selected student is the first one who is involved with the programme;

   ii the fourth selected student is the second one who is involved with the programme;

   iii the first student selected who is involved with the programme will not occur before 3rd selected student;

   iv to select exactly six students involved with the programme we will not need more than 10 selected students?

   b What is the expected number of students needed to be selected if we need six involved in the programme?

4 A continuous random variable X is given by the cumulative distribution function
$$F(x) = \begin{cases} 0, x < 0 \\ \dfrac{x^2}{a^2}, 0 \leq x \leq 2 \\ 1, x > 2 \end{cases}$$

   a Find the positive value of a.

   b Hence determine the probability density function.

   c What is the modal value of the variable X?

5 A random variable X has a probability generating function

$G(t) = \dfrac{t}{a - bt}$, $t \neq \dfrac{a}{b}$, where a and b are nonzero integers.

  a Show that $b = a - 1$.

  b Hence find $E(X)$ in terms of a.

  c Show that $Var(X) = a^2 - a$.

6 a X_1 and X_2 are independent Poisson variables with the same parameter m. Use probability generating functions to show that $X_1 + X_2$ is a Poisson variable and find the parameter.

  b Given that $X_1, X_2, ..., X_n$, $n \in \mathbb{Z}^+$ are independent Poisson random variables having the same parameter m, use mathematical induction to show that $Y = \displaystyle\sum_{k=1}^{n} X_k$ is also a Poisson distribution with the parameter nm.

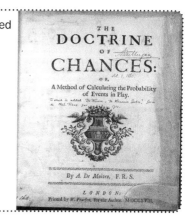

The French mathematician De Moivre (1667–1754) pioneered the development of analytic geometry and the theory of probability. In 1711 he published 'The Doctrine of Chances: A Method of Calculating the Probability of Events in Play' which contained his most significant contribution to this area: the approximation to the binomial distribution by the normal distribution in the case of a large number of trials. De Moivre is famed for predicting the day of his own death. He found that he was sleeping 15 minutes longer each night and, summing the arithmetic progression, calculated that he would die on the day that he slept for 24 hours. He was right!

Chapter 1 summary
Cumulative distribution function

- Given the random variable X (discrete or continuous) and the corresponding probability function $P : \mathbb{R} \to [0, 1]$ the cumulative distribution function $F : \mathbb{R} \to [0, 1]$ is $F(x) = P(X \le x)$, $x \in \mathbb{R}$.

- For continuous random variables the probability density function f and the cumulative distribution function F are related as follows:

$$F'(x) = f(x) \Leftrightarrow F(x) = \int_{-\infty}^{x} f(t)\, dt$$

Geometric distribution

- A discrete random variable X is said to have a geometric distribution and we write $X \sim \text{Geo}(p)$ if $P(X = k) = q^{k-1}p$, where $p \in \mathbb{R}$, $0 \le p \le 1$, $q = 1 - p$, $k = 1, 2, 3, 4 \ldots$

Expected value and variance

- Given a geometric random variable, if $X \sim \text{Geo}(p)$, then $E(X) = \dfrac{1}{p}$ and $\text{Var}(X) = \dfrac{q}{p^2}$.

Negative binomial distribution

- A discrete random variable X is said to have a negative binomial distribution and we write $X \sim \text{NB}(r, p)$ if $P(X = k) = \dbinom{k-1}{r-1} q^{k-r} p^r$, where

$0 \le p \le 1$, $q = 1 - p$, $k = r, r+1, r+2, r+3, \ldots$

Probability generating function

- Let X be a discrete random variable assuming nonnegative integer values and $P(X = k) = p_k$, $k = 0, 1, 2, 3, \ldots$ are the corresponding probabilities. Then the function of the form $G(t) = \sum_{k=0}^{\infty} p_k t^k = p_0 + p_1 t + p_2 t^2 + p_3 t^3 + \ldots + p_n t^n + \ldots$ is called a probability generating function.

Distribution	PGF
Bernoulli $X \sim B(1, p)$	$G(t) = q + pt$
Binomial $X \sim B(n, p)$	$G(t) = (q + pt)^n$
Poisson $X \sim \text{Po}(m)$	$G(t) = e^{m(t-1)}$
Geometric $X \sim \text{Geo}(p)$	$G(t) = \dfrac{pt}{1 - qt}$
Negative binomial $X \sim \text{NB}(r, p)$	$G(t) = \left(\dfrac{pt}{1 - qt} \right)^r$

Properties of PGF and Expected value and Variance

i) $G(1) = 1$

ii) $E(X) = G'(1)$

iii) $\text{Var}(X) = G''(1) + G'(1)(1 - G'(1))$

Independent random variables

- X_1 and X_2 are two independent random variables with the corresponding probability generating functions $G_{X_1}(t)$ and $G_{X_2}(t)$. If a new random variable X is such that $X = X_1 + X_2$ then $G_X(t) = G_{X_1 + X_2}(t) = G_{X_1}(t) \times G_{X_2}(t)$.

Expectation algebra and Central Limit Theorem

CHAPTER OBJECTIVES:

7.2   Linear transformation of a single random variable.
$E(aX + b) = aE(X) + b$, $Var(aX + b) = a^2 Var(X)$
Mean of linear combinations of n random variables.
Variance of linear combinations of n independent random variables.
Expectation of the product of independent random variables. $E(XY) = E(X)E(Y)$

7.4   A linear combination of independent normal random variables is normally
distributed. In particular, the Central Limit Theorem.

Before you start

You should know how to:

1   Given that $X \sim B(5, 0.3)$ find the expected
value and standard deviation of X.

$E(X) = np \Rightarrow E(X) = 5 \times 0.3 = 1.5$

$Var(X) = npq \Rightarrow \sigma = \sqrt{5 \times 0.3 \times 0.7} = 1.02$

2   A random variable X has the
following parameters:

$E(X) = a$ and $Var(X) = 2a^2 - 1$, $a \in \mathbb{R}$.

Find the value of a such that the random
variable follows a Poisson distribution.
The relationship between the expected
value and the variance of a Poisson variable
is $E(X) = Var(X)$ therefore we write

$a = 2a^2 - 1 \Rightarrow 2a^2 - a - 1 = 0$

$(2a + 1)(a - 1) = 0 \Rightarrow a = -\dfrac{1}{2}$ or $a = 1$

Skills check:

1   Given that $X \sim B(n, p)$ with
$E(X) = 2$, $Var(X) = \dfrac{3}{2}$, find the following
probabilities:

a $P(X = 2)$ **b** $P(1 \le X \le 3)$

2   A random variable X has the
following parameters: $E(X) = 7a^2$ and
$Var(X) = 6a^3 - 9a - 2$, $a \in \mathbb{R}$.

Find the value of a such that the random
variable follows a Poisson distribution.

Normal distributions and probability models

In probability theory, expected value refers to the mean value of the set of values obtained when the experiment representing that random variable could be repeated endless number of times. This intuitive explanation of the expected value is a direct consequence of the Law of Large Numbers, i.e. it is the limit of the sample mean as the sample size approaches infinity. This value may not be 'expected' in the English sense of this word, rather it could even be unlikely, or counter-intuitive.

Perhaps some of the most important concepts in the study of a random variable are expected value and variance. We will investigate their algebraic properties in the aspect of so-called expectation algebra. Furthermore, we will be looking at linear combinations of independent random variables and how independence affect the properties of the parameters of such combined random variables.

At the end of this chapter you will study one of the most important results in mathematics, the Central Limit Theorem, and discover the important role that normal distributions have in devising probability models. The expected value and variance are important parameters in probability distribution, and in statistical topics, such as regression analysis, which we will study further in Chapter 4.

> **?** Sometimes, just by chance, a few plane crashes occur in a short period of time. Dramatic headlines and impressive pictures of debris shown in the media may make nervous flyers feel fearful. However, aviation safety data reveals a different reality and a perhaps counter-intuitive truth: air travel is the safest it has been in the history of aviation. The average number of flights that take off each day is very large, and data analysts get a very different picture of reality given by the numbers.

2.1 Expectation algebra

In statistics, measures of central tendency (mean, median, etc.) and measures of dispersion (variance, standard deviation, etc.) change when the values of the data these measures represent are adjusted. In the core companion we conducted two investigations on these changes, and we noted that:

- If a constant k is added to each member of the data set the mean value of the new data set also increased by k;
- If each member of the data set is multiplied by a constant k, the mean value of the new data set is k times the original mean;
- If a constant k is added to each member of the data set the variance of the new data set doesn't change;
- If each member of the data set is multiplied by a constant k, the new variance of the data is k^2 times the original variance.

Since random variables allow us to model random phenomena given by sets of data, we expect that the mean value and variance of the set of data will behave similarly to the expected value and variance of a random variable.

> Linear transformation of a variable is a transformation that is obtained by addition and multiplication of the variable by constants.

Linear transformation of a single variable

Example 1

Luca and his father play a game. Luca flips a coin. If he gets a "head" his father will give him €1.

a Find the expected value and the variance of the random variable X that describes the amount of money Luca will get in this game.

Luca's father later decides to award him with €5 if he obtains a "head".

b Find the expected value and the variance of the random variable Y that describes the amount of money Luca will get in this new version of the game.

c Suggest a relationship between the expected values and the variances of the variables X and Y.

---

a

$X = x_i$	0	1
$P(X = x_i)$	$\dfrac{1}{2}$	$\dfrac{1}{2}$

$$E(X) = \sum_{i=1}^{2} x_i p_i$$

$$E(X) = 0 \times \frac{1}{2} + 1 \times \frac{1}{2} = \frac{1}{2}$$

$$Var(X) = \sum_{i=1}^{2} x_i^2 p_i - (E(X))^2$$

$$Var(X) = \left(0^2 \times \frac{1}{2} + 1^2 \times \frac{1}{2}\right) - \left(\frac{1}{2}\right)^2 = \frac{1}{2} - \frac{1}{4} = \frac{1}{4}$$

Luca gets nothing if he obtains a "tail" and gets €1 if he obtains a "head". Draw the probability distribution table for the variable X and fill in all the values.

Calculate the expected value.

Calculate the variance.

b

$Y = y_i$	0	5
$P(Y = y_i)$	$\dfrac{1}{2}$	$\dfrac{1}{2}$

Luca gets nothing if he obtains a "tail" and gets €5 if he obtains a "head". Draw the probability distribution table for the variable Y and fill in all the values.

$$E(Y) = 0 \times \frac{1}{2} + 5 \times \frac{1}{2} = \frac{5}{2}$$

Calculate the expected value.

$$Var(Y) = \left(0^2 \times \frac{1}{2} + 5^2 \times \frac{1}{2}\right) - \left(\frac{5}{2}\right)^2 = \frac{25}{2} - \frac{25}{4} = \frac{25}{4}$$

Calculate the variance.

c $\quad E(Y) = 5E(X)$

$\quad\quad Var(Y) = 25\, Var(X)$

By looking at the possible values of the variables we notice that $y_i = 5x_i$, $i = 1, 2$ so we write $Y = 5X$. Notice that the expected value was increased by the same factor and the variance was increased by the square of the factor.

Now let's consider another variation of the game.

Example 2

Luca and his father again play a similar game. Luca's father will give him €1 if he obtains a "head" but regardless of the outcome obtained he will give him an extra €2.

a Find the expected value and the variance of the random variable Z that describes the amount of money Luca will get in this game.

b Suggest a relationship between the expected values and the variances of the variables X (in Example 1), and Z.

---

a

$Z = z_i$	2	3
$P(Z = z_i)$	$\dfrac{1}{2}$	$\dfrac{1}{2}$

Luca gets €2 if he obtains a "tail" and gets €3 if he obtains a "head". Draw the probability distribution table for the variable Z and fill in all the values.

$$E(Z) = 2 \times \frac{1}{2} + 3 \times \frac{1}{2} = \frac{5}{2}$$

Calculate the expected value.

$$Var(Z) = \left(2^2 \times \frac{1}{2} + 3^2 \times \frac{1}{2}\right) - \left(\frac{5}{2}\right)^2 = \frac{13}{2} - \frac{25}{4} = \frac{1}{4}$$

Calculate the variance.

b $\quad E(Z) = \dfrac{5}{2} = \dfrac{1}{2} + 2 = E(X) + 2$

We can see that $z_i = x_i + 2$, $i = 1, 2$ so we write $Z = X + 2$.

$\quad\quad Var(Z) = \dfrac{1}{4} = Var(X)$

Notice that the expected value was increased by the same added value while the variance remained the same.

Example 3 looks at one last variation of the game.

Example 3

Luca's father said that he will give Luca €10 if he gets a "head" when flipping a coin.
a Find the expected value and the variance of the random variable R that describes the amount of money Luca will get in the game.

Luca's father then decides to make it more realistic and introduces a €3 tax whenever Luca flips a coin.
b Find the expected value and the variance of the random variable S that describes the amount of money Luca will get in this variation of the game.
c Suggest a relationship between the expected values and the variances of the variables R and S.

---

a

$R = r_i$	0	10
$P(R = r_i)$	$\dfrac{1}{2}$	$\dfrac{1}{2}$

Luca gets nothing if he obtains a "tail" and gets €10 if he obtains a "head". Draw the probability distribution table for the variable R and fill in all the values.

$$E(R) = 0 \times \frac{1}{2} + 10 \times \frac{1}{2} = 5$$

Calculate the expected value.

$$\text{Var}(R) = \left(0^2 \times \frac{1}{2} + 10^2 \times \frac{1}{2}\right) - 5^2 = 50 - 25 = 25$$

Calculate the variance.

b

$S = s_i$	−3	7
$P(S = s_i)$	$\dfrac{1}{2}$	$\dfrac{1}{2}$

Luca loses €3 if he obtains a "tail" and gets €7 (€10 – €3) if he obtains a "head". Draw the probability distribution table for the variable S and fill in all the values.

$$E(S) = -3 \times \frac{1}{2} + 7 \times \frac{1}{2} = 2$$

Calculate the expected value.

$$\text{Var}(S) = \left((-3)^2 \times \frac{1}{2} + 7^2 \times \frac{1}{2}\right) - 2^2 = \frac{58}{2} - 4 = 25$$

Calculate the variance.

c $E(S) = 2 = 5 - 3 = E(R) - 3$
$\text{Var}(S) = 25 = \text{Var}(R)$

We can see that $s_i = r_i - 3$, $i = 1, 2$ so we can write $S = R - 3$.

Notice that the expected value was reduced by the same value that was subtracted from the variable, whilst the variance remains the same.

These examples suggest that all the changes from the data adjustment are valid for random variables too. The following theorem formalizes these results about the parameters of a linear transformation of a random variable.

Notice that the probability of a random variable taking a particular outcome remains the same when we perform a linear transformation on the random variable.

Theorem 1

Given that X is a random variable with finite parameters and $a, b \in \mathbb{R}$, then

i $E(aX + b) = aE(X) + b$;

ii $\text{Var}(aX + b) = a^2\text{Var}(X)$.

Proof:

As random variables can be either discrete or continuous, we must prove the theorem for both cases:

Case I

Let X be a discrete random variable. Then:

i $\mu = E(X) = \sum xP(X = x) \Rightarrow$

$$E(aX + b) = \sum(ax + b)P(X = x) = \sum(axP(X = x) + bP(X = x))$$

$$= \sum axP(X = x) + \sum bP(X = x) = a\underbrace{\sum xP(X = x)}_{E(X)} + b\underbrace{\sum P(X = x)}_{1}$$

$$= aE(X) + b$$

$$\boxed{\text{Var}(X) = E(X^2) - (E(X))^2, E(X) = \mu}$$

ii $\text{Var}(X) = \sum x^2 P(X = x) - \mu^2 \Rightarrow$

$$\text{Var}(aX + b) = \sum(ax + b)^2 P(X = x) - (a\mu + b)^2$$

$$= \sum(a^2x^2 + 2axb + b^2)P(X = x) - (a^2\mu^2 + 2a\mu b + b^2)$$

$$= \sum(a^2x^2P(X = x) + 2axbP(X = x) + b^2 P(X = x)) - (a^2\mu^2 + 2a\mu b + b^2)$$

$$= a^2\sum x^2 P(X = x) + 2ab\underbrace{\sum xP(X = x)}_{E(X)} + b^2\underbrace{\sum P(X = x)}_{1} - (a^2\mu^2 + 2a\mu b + b^2)$$

$$= a^2\sum x^2 P(X = x) + 2ab\mu + b^2 - a^2\mu^2 - 2a\mu b - b^2$$

$$= a^2\left(\underbrace{\sum x^2P(X = x) - \mu^2}_{\text{Var}(X)}\right) = a^2\,\text{Var}(X)$$

Case II Let X be a continuous random variable. Then:

i $\mu = E(X) = \int xf(x)\,dx \Rightarrow$

$$E(aX + b) = \int(ax + b)f(x)dx = \int(axf(x) + bf(x))dx = a\underbrace{\int xf(x)dx}_{E(X)} + b\underbrace{\int f(x)dx}_{1}$$

$$= aE(X) + b$$

ii $\text{Var}(X) = \int x^2 f(x)dx - \mu^2 \Rightarrow$

$$\text{Var}(aX + b) = \int (ax + b)^2 f(x)dx - (a\mu + b)^2$$

$$= \int (a^2x^2 + 2abx + b^2)f(x)dx - (a^2\mu^2 + 2ab\mu + b^2)$$

$$= a^2 \int x^2 f(x)dx + 2ab \underbrace{\int xf(x)dx}_{E(X)} + b^2 \underbrace{\int f(x)dx}_{1} - (a^2\mu^2 + 2ab\mu + b^2)$$

$$= a^2 \int x^2 f(x)dx + \cancel{2ab\mu} + \cancel{b^2} - a^2\mu^2 - \cancel{2ab\mu} - \cancel{b^2}$$

$$= a^2 \underbrace{\left(\int x^2 f(x)dx - \mu^2 \right)}_{\text{Var}(X)}$$

$$= a^2 \text{Var}(X)$$

Exercise 2A

1 Given that X is a random variable with the expected value 5.3 and variance 1.2, find the expected value and variance of the following:

    a $3X$ **b** $X + 3$ **c** $4X + 1$ **d** $2X - 5$ **e** $kX + p;\ k,\ p \in \mathbb{R}$

2 Given that a random variable $X \sim B\left(10, \dfrac{2}{5}\right)$ find:

    a $E(3X + 2)$ **b** $\text{Var}(3X - 2)$

3 A random variable Y follows a geometric distribution with the parameter $p = \dfrac{2}{3}$. Find the expected value and variance of $2Y - 1$.

4 Given that a random variable $Y \sim \text{Po}(2)$ find:

    a $E(3 - 2Y)$ **b** $\text{Var}(3 - 2Y)$

5 Given that a random variable $X \sim \text{NB}\left(8, \dfrac{1}{3}\right)$ find:

    a $E(2X - 3)$ **b** $\text{Var}(2X - 11)$

6 Given that a random variable $X \sim B(15, p)$ and $E(X) = 6$, find $\text{Var}(5X + 3)$.

7 A continuous random variable X has a probability density function given by the formula $f(x) = \begin{cases} \dfrac{1}{3}x, & 0 \le x \le \sqrt{6} \\ 0, & \text{otherwise} \end{cases}$

Find the exact values of the expected value and variance of $3X + 2$.

So far we were performing a linear transformation of just one variable.
Let's see what happens if we involve two or more random variables.
We are going to assume that the variables are independent.

Linear transformation of two or more variables

Example 4

Hannah and Luca play a game with their father by flipping a coin. Hannah will get €2 if she obtains a "head", whilst Luca will get €5 if he obtains a "head". Random variables X and Y represent the amount of money Hannah and Luca get in this game respectively. The random variable Z represents the amount they earn together in this game.

a Draw the probability distribution table and calculate the expected value and variance of Z.

Their grandmother decides to increase the amount they will get. Hannah will get 8 times more money than she received from her father, and Luca will get 3 times more money than he received from his father. Again they put their money together.

b Find the probability distribution of the random variable W that describes the amount of money they will earn together this time. Calculate the expected value and variance of W.

c Suggest relationships between the expected values and the variances of the variables X, Y, Z and W.

---

a

$X = x_i$	0	2
$P(X = x_i)$	$\dfrac{1}{2}$	$\dfrac{1}{2}$

$E(X) = 1, \quad \mathrm{Var}(X) = 1$

$Y = y_i$	0	5
$P(Y = y_i)$	$\dfrac{1}{2}$	$\dfrac{1}{2}$

$E(Y) = \dfrac{5}{2}, \quad \mathrm{Var}(Y) = \dfrac{25}{4}$

$Z = X + Y$

$Z = z_i$	0	2	5	7
$P(Z = z_i)$	$\dfrac{1}{4}$	$\dfrac{1}{4}$	$\dfrac{1}{4}$	$\dfrac{1}{4}$

$$E(Z) = 0 \times \frac{1}{4} + 2 \times \frac{1}{4} + 5 \times \frac{1}{4} + 7 \times \frac{1}{4} = \frac{7}{2}$$

$$\mathrm{Var}(Z) = \left(\frac{0^2}{4} + \frac{2^2}{4} + \frac{5^2}{4} + \frac{7^2}{4} \right) - \left(\frac{7}{2} \right)^2$$

$$= \frac{78}{4} - \frac{49}{4} = \frac{29}{4}$$

Hannah gets nothing if she obtains a "tail" and gets €2 if she obtains a "head". Draw the probability distribution table for the variable X and fill in all the values.

Calculate expected value and variance. Luca gets nothing if he obtains a "tail" and gets €5 if he obtains a "head". Draw the probability distribution table for the variable Y and fill in all the values.

Calculate expected value and variance.

If they put the money together then we are adding the money they earn.

The values of Z for the corresponding pairs of outcomes:
(T, T) → 0 + 0, (H, T) → 2 + 0
(T, H) → 0 + 5, (H, H) → 2 + 5
Calculate the expected value.

Calculate the variance.

b $W = 8X + 3Y$

$W = w_i$	0	15	16	31
$P(W = w_i)$	$\dfrac{1}{4}$	$\dfrac{1}{4}$	$\dfrac{1}{4}$	$\dfrac{1}{4}$

Again we are adding the money they earn. The values of W for the corresponding pairs of outcomes:
$(T, T) \rightarrow 0 + 0$, $(T, H) \rightarrow 0 + 3 \times 5$
$(H, T) \rightarrow 8 \times 2 + 0$,
$(H, H) \rightarrow 8 \times 2 + 3 \times 5$

$$E(W) = 0 \times \frac{1}{4} + 15 \times \frac{1}{4} + 16 \times \frac{1}{4} + 31 \times \frac{1}{4} = \frac{31}{2}$$

Calculate the expected value.

$$Var(X) = \left(\frac{0^2}{4} + \frac{15^2}{4} + \frac{16^2}{4} + \frac{31^2}{4}\right) - \left(\frac{31}{2}\right)^2$$

Calculate the variance.

$$= \frac{721}{2} - \frac{961}{4} = \frac{481}{4}$$

c $E(X) + E(Y) = E(X + Y) = E(Z)$

Notice that $1 + \dfrac{5}{2} = \dfrac{7}{2}$.

$Var(X) + Var(Y) = Var(X + Y) = Var(Z)$

Notice that $1 + \dfrac{25}{4} = \dfrac{29}{4}$.

$8E(X) + 3E(Y) = E(8X + 3Y) = E(W)$

Notice that $8 \times 1 + 3 \times \dfrac{5}{2} = \dfrac{31}{2}$.

$64Var(X) + 9Var(Y) = Var(8X + 3Y) = Var(W)$

Notice that $64 \times 1 + 9 \times \dfrac{25}{4} = \dfrac{481}{4}$.

Now let's take another example, this time with probabilities that are not uniformly distributed.

Example 5

Mia and Theo draw marbles from two boxes. Mia draws a marble from the first box; it contains 2 white and 3 black marbles. Theo draws two marbles form the second box; it contains 4 white and 3 black marbles. Random variables X and Y represent the number of white marbles Mia and Theo are going to draw, respectively. The random variable Z represents the number of white marbles they draw together.

a Draw the probability distribution tables and calculate the expected value and the variance of all the variables X, Y and Z.

Their grandfather decides to give Mia €3 and Theo €2 for each white marble they draw. Mia and Theo decide to put the money together.

b Find the probability distribution of the random variable W that describes the amount of money they earn together, and calculate the expected value and variance of W.

c Suggest a relationship between the expected values and variances of the variables X, Y, Z and W.

a

$X = x_i$	0	1
$P(X = x_i)$	$\dfrac{3}{5}$	$\dfrac{2}{5}$

Mia can draw one black or one white marble from the first box of marbles. Draw the probability distribution table for the variable X and fill in all the values. Calculate expected value and variance.

$$E(X) = \frac{2}{5}, \quad Var(X) = \frac{2}{5} - \left(\frac{2}{5}\right)^2 = \frac{6}{25}$$

$$P(Y=0) = \frac{\binom{3}{2}}{\binom{7}{2}} = \frac{1}{7}, \quad P(Y=1) = \frac{\binom{3}{1}\binom{4}{1}}{\binom{7}{2}} = \frac{4}{7},$$

$$P(Y=2) = \frac{\binom{4}{2}}{\binom{7}{2}} = \frac{2}{7}$$

$Y = y_i$	0	1	2
$P(Y = y_i)$	$\frac{1}{7}$	$\frac{4}{7}$	$\frac{2}{7}$

$$E(Y) = \frac{8}{7}, \quad Var(Y) = \left(\frac{4}{7} + \frac{8}{7}\right) - \left(\frac{8}{7}\right)^2 = \frac{20}{49}$$

$$Z = X + Y$$

$$P(Z=0) = \frac{3}{5} \times \frac{1}{7} = \frac{3}{35}$$

$$P(Z=1) = \frac{2}{5} \times \frac{1}{7} + \frac{3}{5} \times \frac{4}{7} = \frac{14}{35} = \frac{2}{5}$$

$$P(Z=2) = \frac{2}{5} \times \frac{4}{7} + \frac{3}{5} \times \frac{2}{7} = \frac{14}{35} = \frac{2}{5}$$

$$P(Z=3) = \frac{2}{5} \times \frac{2}{7} = \frac{4}{35}$$

$Z = z_i$	0	1	2	3
$P(Z = z_i)$	$\frac{3}{35}$	$\frac{2}{5}$	$\frac{2}{5}$	$\frac{4}{35}$

$$E(Z) = 0 \times \frac{3}{35} + 1 \times \frac{2}{5} + 2 \times \frac{2}{5} + 3 \times \frac{4}{35} = \frac{54}{35}$$

$$Var(Z) = \left(0 + \frac{2}{5} + \frac{8}{5} + \frac{36}{35}\right) - \left(\frac{54}{35}\right)^2$$

$$= \frac{106}{35} - \frac{2916}{1225} = \frac{794}{1225}$$

b $W = 3X + 2Y$

$W = w_i$	0	2	3	4	5	7
$P(W = w_i)$	$\frac{3}{35}$	$\frac{12}{35}$	$\frac{2}{35}$	$\frac{6}{35}$	$\frac{8}{35}$	$\frac{4}{35}$

Theo can draw two black marbles or one black and one white or two white marbles from the second box of marbles.

Draw the probability distribution table for the variable Y and fill in all the values.

Calculate the expected value and variance.

If they put the marbles together then we are actually adding the number of marbles they drew. $(Z = 0) \leftrightarrow (0 + 0)$

$(Z = 1) \leftrightarrow (1 + 0)$ or $(0 + 1)$

$(Z = 2) \leftrightarrow (1 + 1)$ or $(0 + 2)$

$(Z = 3) \leftrightarrow (1 + 2)$

Draw the probability distribution table for the variable Z and fill in all the values.

Calculate the expected value.

Calculate the variance.

This time we are adding the money they earn, not the number of white marbles they draw. The values of W for the corresponding pairs of outcomes:
$0 + 0, 0 + 2, 3 + 0, 0 + 4, 3 + 2, 3 + 4$

$$E(W) = 0 + \frac{24}{35} + \frac{6}{35} + \frac{24}{35} + \frac{40}{35} + \frac{28}{35} = \frac{122}{35}$$

Calculate the expected value.

$$\text{Var}(X) = \left(0 + \frac{48}{35} + \frac{18}{35} + \frac{96}{35} + \frac{200}{35} + \frac{196}{35}\right) - \left(\frac{122}{35}\right)^2$$

Calculate the variance.

$$= \frac{558}{35} - \frac{14{,}884}{1225} = \frac{4646}{1225}$$

c $E(X) + E(Y) = E(X+Y) = E(Z)$

Notice that $\dfrac{2}{5} + \dfrac{8}{7} = \dfrac{54}{35}$

$\text{Var}(X) + \text{Var}(Y) = \text{Var}(X+Y) = \text{Var}(Z)$

Notice that $\dfrac{6}{25} + \dfrac{20}{49} = \dfrac{794}{1225}$

$3E(X) + 2E(Y) = E(3X+2Y) = E(W)$

Notice that $3 \times \dfrac{2}{5} + 2 \times \dfrac{8}{7} = \dfrac{122}{35}$

$9\text{Var}(X) + 4\text{Var}(Y) = \text{Var}(3X+2Y) = \text{Var}(W)$

Notice that $9 \times \dfrac{6}{25} + 4 \times \dfrac{20}{49} = \dfrac{4646}{1225}$

Examples 4 and 5 demonstrated one very important property for the probability of the sum of two independent random variables. To find the probability of the variable $Z = X + Y$ we multiplied the corresponding elementary probabilities of the random variables X and Y:

$$P(Z = z) = P(X = x) \times P(Y = y)$$

In the second part of Examples 4 and 5, we showed that the probability of a variable W (which was a linear combination of X and Y involving real coefficients, $W = 3X + 2Y$) was simply the product of the probability of X and Y, and is totally independent of the coefficients:

$$P(W = w) = P(X = x) \times P(Y = y)$$

We also demonstrated that the expected value and variance of linear combinations of multiple random variables behave in the same way as the expected value and variance of linear combinations of a single random variable. We can now state the following theorem.

> **?** $E(X)$ is also called the first moment of X. The second moment, $E(X - X)^2$ is also called the variance of X which measures the spread and relates to variability, volatility, and uncertainty. We can also define the nth moment: $E(X - X)^n$ For example, the third moment: $E(X - X)^3$ is a measure of skewness or asymmetry in the distribution of X.

Theorem 2

Given that X and Y are two independent random variables with finite parameters and $a, b \in \mathbb{R}$, then:

i $E(aX + bY) = aE(X) + bE(Y)$

ii $\text{Var}(aX + bY) = a^2\text{Var}(X) + b^2\text{Var}(Y)$

Proof:

We are going to prove this theorem for a discrete random variable only:

i $\mu_1 = \mathrm{E}(X) = \sum_x x\mathrm{P}(X=x), \quad \mu_2 = \mathrm{E}(Y) = \sum_y y\mathrm{P}(Y=y)$	*Since X and Y are independent variables the probability of the intersection is the product of probabilities.*
$\Rightarrow \mathrm{E}(aX+bY) = \sum_{x,y}(ax+by)\mathrm{P}((X=x)\cap(Y=y))$	
$= \sum_{x,y}((ax+by)\mathrm{P}(X=x)\mathrm{P}(Y=y))$	
$= \sum_x\sum_y (ax\,\mathrm{P}(X=x)\mathrm{P}(Y=y) + by\,\mathrm{P}(X=x)\mathrm{P}(Y=y))$	*Use the distributive property.*
$= \underbrace{\sum_y \mathrm{P}(Y=y)}_{1}\left(\sum_x ax\,\mathrm{P}(X=x)\right)$	*The sum of all the probabilities of a random variable is equal to 1.*
$+ \underbrace{\sum_x \mathrm{P}(X=x)}_{1}\left(\sum_y by\,\mathrm{P}(Y=y)\right)$	
$= a\underbrace{\sum_x x\mathrm{P}(X=x)}_{\mathrm{E}(X)} + b\underbrace{\sum_y y\mathrm{P}(Y=y)}_{\mathrm{E}(Y)}$	*Use the definition of the expected value.*
$= a\mathrm{E}(X) + b\mathrm{E}(Y)$	
ii $\mathrm{Var}(X) = \sum_x (x^2\mathrm{P}(X=x)) - \mu_1^2$	
$\mathrm{Var}(Y) = \sum_y (y^2\mathrm{P}(Y=y)) - \mu_2^2 \Rightarrow$	
$\mathrm{Var}(aX+bY) = \sum_{x,y}(ax+by)^2\mathrm{P}((X=x)\cap(Y=y))$	*Since X and Y are independent variables, the probability of the intersection is the product of probabilities.*
$-(a\mu_1+b\mu_2)^2$	
$= \sum_{x,y}((a^2x^2 + 2axby + b^2y^2)\,\mathrm{P}(X=x)\mathrm{P}(Y=y))$	
$-(a^2\mu_1^2 + 2a\mu_1 b\mu_2 + b^2\mu_2^2)$	*Use the distributive property. The sum of all the probabilities of a random variable is equal to 1. Recognize the expression for the expected value.*
$= \underbrace{\sum_y \mathrm{P}(Y=y)}_{1}\sum_x (a^2x^2\,\mathrm{P}(X=x))$	
$+ 2ab\underbrace{\sum_x (x\mathrm{P}(X=x))}_{\mu_1}\underbrace{\sum_y (y\mathrm{P}(Y=y))}_{\mu_2}$	
$+ \underbrace{\sum_x \mathrm{P}(X=x)}_{1}\sum_y (b^2y^2\mathrm{P}(Y=y))$	
$-(a^2\mu_1^2 + 2a\mu_1 b\mu_2 + b^2\mu_2^2)$	*Reduce opposite terms and collect the like terms.*

$$= a^2 \sum_x x^2 P(X = x) \underline{+2ab\mu_1\mu_2} + b^2 \sum_y (y^2 P(Y = y))$$

$$\underline{-a^2\mu_1^2} \quad \underline{-2a\mu_1 b\mu_2} \quad \underline{-b^2\mu_2^2}$$

$$= a^2 \left(\sum_x (x^2 P(X = x)) - \mu_1^2 \right) + b^2 \left(\sum_y (y^2 P(Y = y)) - \mu_2^2 \right)$$

$$= a^2 \text{Var}(X) + b^2 \text{Var}(Y)$$

Use distributive property and the definition of the variance.

The proof of the theorem for a continuous random variable is beyond the scope of this syllabus.

Theorem 2 can now be generalised to give a result for a finite number of random variables.

Theorem 3

Given that $X_1, X_2, X_3, ..., X_n$, $n \in \mathbb{Z}^+$ are independent random variables with finite parameters and $a_1, a_2, a_3, ... a_n \in \mathbb{R}$ then:

i $\text{E}(a_1 X_1 + a_2 X_2 + ... + a_n X_n) = a_1 \text{E}(X_1) + a_2 \text{E}(X_2) + ... + a_n \text{E}(X_n);$

ii $\text{Var}(a_1 X_1 + a_2 X_2 + ... + a_n X_n)$
$= a_1^2 \text{Var}(X_1) + a_2^2 \text{Var}(X_2) + ... + a_n^2 \text{Var}(X_n).$

Using sigma notation we can write:

i $\text{E}\left(\sum_{i=1}^{n} a_i X_i \right) = \sum_{i=1}^{n} a_i \text{E}(X_i)$

ii $\text{Var}\left(\sum_{i=1}^{n} a_i X_i \right) = \sum_{i=1}^{n} a_i^2 \text{Var}(X_i)$

Unlike in the calculation of expected value, we notice that when calculating variance it doesn't matter whether the coefficients a and b are positive or negative; their squares are always positive. In particular we have to be careful that
$\text{E}(X \pm Y) = \text{E}(X) \pm \text{E}(Y)$ and
$\text{Var}(X \pm Y) = \text{Var}(X) + \text{Var}(Y)$

The proof of this theorem for the discrete case can be done by mathematical induction and is left as an exercise for students.

Exercise 2B

1 The random variables X, Y and Z are given with their parameters in the table below.

Variable	μ	σ^2
X	3	0.5
Y	-5	1.4
Z	12	2.8

Find the parameters of the following linear combinations:

a $X + Y$ **b** $2Y - Z$ **c** $2Z - 7X$

d $X - Y + Z$ **e** $X + Y - Z$ **f** $3Z - 2X + 4Y$

2 Poisson random variables X and Y are such that $\text{Var}(X) = 2$ and $\text{E}(Y) = 5$. Calculate:

    a $\text{E}(3X + 5Y)$ **b** $\text{Var}(11Y - 7X)$

3 Binomial random variables X and Y are such that $\text{E}(X) = 9$ and $\text{E}(Y) = 4$. Given that they have the same probability of success, p, find the variance of $2X - 3Y$ in terms of p.

4 Negative binomial random variables X and Y are such that $\text{E}(X) = 8$ and $\text{E}(Y) = 12$. Given that they have the same probability of success, p, find the variance of $X - Y$ in terms of p.

5 Use mathematical induction to prove Theorem 3.

In the manipulation of random variables we have explored so far, we haven't yet discussed whether a linear combination of variables which are all distributed in the same way (e.g. all have Poisson distribution) will give a random variable that is also distributed in this way.

Example 6

Given two Poisson random variables $X \sim \text{Po}(\mu_1)$ and $Y \sim \text{Po}(\mu_2)$, find the expected value and the variance of the random variables:

a $X + Y$

b $X - Y$

c $4X - 3Y$

a $\text{E}(X + Y) = \text{E}(X) + \text{E}(Y) = \mu_1 + \mu_2$ $\text{Var}(X + Y) = \text{Var}(X) + \text{Var}(Y) = \mu_1 + \mu_2$	*Both the expected value and variance of a Poisson random variable are equal to the parameter. Apply Theorem 2. Notice that the expected value of the random variable $X + Y$ is equal to the variance. Hence, we might speculate that $X + Y$ could also be Poisson.*
b $\text{E}(X - Y) = \text{E}(X) - \text{E}(Y) = \mu_1 - \mu_2$ $\text{Var}(X - Y) = \text{Var}(X) + \text{Var}(Y) = \mu_1 + \mu_2$	*Again apply Theorem 2. Notice that the expected value of the random variable $X - Y$ is not equal to its variance. Thus, $X - Y$ is not Poisson.*
c $\text{E}(4X - 3Y) = 4\text{E}(X) - 3\text{E}(Y) = 4\mu_1 - 3\mu_2$ $\text{Var}(4X - 3Y) = 4^2\text{Var}(X) + (-3)^2\text{Var}(Y)$ $= 16\mu_1 + 9\mu_2$	*Again apply Theorem 2 and notice that the expected value of the random variable $4X - 3Y$ is not equal to its variance. Thus, $4X - 3Y$ is not Poisson.*

We can conclude that, in general, a linear combination of two Poisson random variables is not a Poisson random variable.

In a similar situation we ask ourselves what is going to happen if two random variables have equal expected values $\mu_1 = \mu_2$ and equal variances $\sigma_1^2 = \sigma_2^2$. In this case we conduct a slightly different calculation.

In the following example, we will use the expected value and variance of a random variable X to investigate whether or not we can conclude that $2X = X + X$.

Example 7

Mia draws one marble from a box. She notes the colour of the marble, and then returns it back to the box. The box contains 2 white and 3 black marbles.

a Random variable X represents the number of white marbles Mia draws.
   Find the expected value and the variance of X.

b Mia's uncle says that he will give €2 to Mia for each white marble she draws.
   The random variable Y denotes the amount of money Mia receives from her uncle.
   Find the expected value and variance of Y.

Mia's brother, Theo, draws a marble from the same box and notes the colour. He then replaces it. Afterwards, Mia draws a marble from the box, notes the colour and replaces it.

c Their aunt says she will give them €1 for each white marble drawn. The random variable Z denotes the amount of money they will get from their aunt. Find the expected value and variance of Z.

d Given that we can write $Y = 2X$ and $Z = X + X$, what do you notice about the parameters of the variables Y and Z?

a

$X = x_i$	0	1
$P(X = x_i)$	$\dfrac{3}{5}$	$\dfrac{2}{5}$

The probability distribution table for the variable X is the same as that in Example 5.

$$E(X) = \frac{2}{5}, \quad Var(X) = \frac{2}{5} - \left(\frac{2}{5}\right)^2 = \frac{6}{25}$$

Calculate the expected value and variance.

b

$Y = y_i$	0	2
$P(Y = y_i)$	$\dfrac{3}{5}$	$\dfrac{2}{5}$

Draw the probability distribution table for the variable Y and fill in the probabilities.

$$E(X) = \frac{4}{5}, \quad Var(X) = \frac{8}{5} - \left(\frac{4}{5}\right)^2 = \frac{24}{25}$$

Calculate the expected value and variance.

c

$Z = z_i$	0	1	2
$P(Z = z_i)$	$\dfrac{9}{25}$	$\dfrac{12}{25}$	$\dfrac{4}{25}$

Draw the probability distribution table for the variable Z and fill in all the values. We are actually adding the number of marbles they draw.

$$P(Z = 0) = \frac{3}{5} \times \frac{3}{5} = \frac{9}{25}$$

(Z = 0) ↔ (0 + 0)

$$P(Z = 1) = \frac{2}{5} \times \frac{3}{5} + \frac{3}{5} \times \frac{2}{5} = \frac{12}{25}$$

(Z = 1) ↔ (1 + 0) or (0 + 1)

$$P(Z = 2) = \frac{2}{5} \times \frac{2}{5} = \frac{4}{25}$$

(Z = 2) ↔ (1 + 1)

$$E(Z) = 0 \times \frac{3}{25} + 1 \times \frac{12}{25} + 2 \times \frac{4}{5} = \frac{20}{25} = \frac{4}{5}$$

Calculate the expected value and variance.

$$Var(Z) = \left(0 + \frac{12}{25} + \frac{16}{25}\right) - \left(\frac{4}{5}\right)^2 = \frac{12}{25}$$

Notice that Var(Z) = Var(X) + Var(X).

d
$$E(Y) = E(Z)$$
$$E(2X) = E(X + X)$$
$$Var(2X) \neq Var(X + X)$$

Notice that although X and Z have equal expected values, they don't have equal variances. Thus, we conclude that for random variables in general $2X \neq X + X$.

Theorem 4 will generalize the conclusion we reached in Example 7:

Theorem 4

Given that independent random variables $X_1, X_2, X_3, ..., X_n, n \in \mathbb{Z}^+$

all have equal expected value and equal variance $E(X) = \mu$ and $Var(X) = \sigma^2$, then:

$$E(X_1 + X_2 + ... + X_n) = E(X_1) + E(X_2) + ... + E(X_n) = \underbrace{\mu + \mu + ... + \mu}_{n \text{ terms}} = n\mu$$

and

$$Var(X_1 + X_2 + ... + X_n) = Var(X_1) + Var(X_2) + ... + Var(X_n) = \underbrace{\sigma^2 + \sigma^2 + ... + \sigma^2}_{n \text{ terms}} = n\sigma^2$$

When adding n random variables which have the same expected value, the expectation of the sum $(X_1 + X_2 + ... + X_n)$ is **equal** to
the expectation of the variable multiplied by n, (nX), i.e.
$E(X_1 + X_2 + ... + X_n) = E(nX)$. The variance of the sum **is not equal** to the variance in the case of a linear transformation of a random variable,

$$Var(X_1 + X_2 + ... + X_n) = n\sigma^2 \neq n^2\sigma^2 = Var(nX).$$

Therefore, by comparing parameters, we conclude that for a random variable X:

$$\underbrace{X + X + ... + X}_{n \text{ terms}} \neq nX.$$

Example 8

Anna and Sarah are independently drawing counters from two different boxes.
Anna's box contains the counters 1, 2, 3, 4 and 5. Sarah's box contains the counters
1, 1, 1, 2, 2 and 3. Let random variables X and Y represent the value of the counters
drawn by Anna and Sarah respectively.

a Draw the probability distribution tables and find the expected values of the
   variables X and Y.

Anna's father is going to give Anna as much money as the value of her drawn counter
multiplied by the value of Sarah's drawn counter. Let random variable Z represent the
amount of money Anna will earn in this game.

b Draw the probability distribution table and find the expected value of the variable Z.
c Suggest a relationship between the variables X, Y and Z.

---

a

$X = x_i$	1	2	3	4	5
$P(X = x_i)$	$\dfrac{1}{5}$	$\dfrac{1}{5}$	$\dfrac{1}{5}$	$\dfrac{1}{5}$	$\dfrac{1}{5}$

Notice that X follows a uniform distribution since we have one counter in each value.

$$E(X) = \frac{1}{5} \times (1+2+3+4+5) = \frac{1}{5} \times 15 = 3$$

Calculate the expected value.

$Y = y_i$	1	2	3
$P(Y = y_i)$	$\dfrac{1}{2}$	$\dfrac{1}{3}$	$\dfrac{1}{6}$

There are only three different values on the counters and there are 3 out of 6 counters with the value "1", 2 out of 6 counters with the value "2" and only 1 out of 6 counters with the value "3".

$$E(Y) = \frac{1}{2} \times 1 + \frac{1}{3} \times 2 + \frac{1}{6} \times 3 = \frac{5}{3}$$

Calculate the expected value.

b

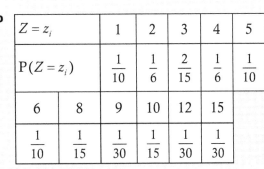

$Z = z_i$	1	2	3	4	5
$P(Z = z_i)$	$\dfrac{1}{10}$	$\dfrac{1}{6}$	$\dfrac{2}{15}$	$\dfrac{1}{6}$	$\dfrac{1}{10}$
6	8	9	10	12	15
$\dfrac{1}{10}$	$\dfrac{1}{15}$	$\dfrac{1}{30}$	$\dfrac{1}{15}$	$\dfrac{1}{30}$	$\dfrac{1}{30}$

We need to look at all the possible pairs and their products. There are altogether 11 pairs of outcomes and since the drawings are independent for each of them we have to multiply their probabilities. Some of the pairs give the same product so we have to add their probabilities. For example:

$$1 : (1,\ 1) \rightarrow \frac{1}{5} \times \frac{1}{2} = \frac{1}{10}$$

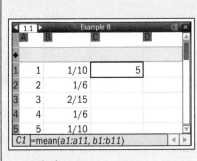

$$2 : (1, 2) \text{ or } (2, 1) \rightarrow \frac{1}{5} \times \frac{1}{3} + \frac{1}{5} \times \frac{1}{2} = \frac{1}{6}$$

$$3 : (1, 3) \text{ or } (3, 1) \rightarrow \frac{1}{5} \times \frac{1}{6} + \frac{1}{5} \times \frac{1}{2} = \frac{2}{15}$$

$$4 : (2, 2) \text{ or } (4, 1) \rightarrow \frac{1}{5} \times \frac{1}{3} + \frac{1}{5} \times \frac{1}{2} = \frac{1}{6}$$

To simplify the calculation for expectation we can use a GDC.

$$\mathrm{E}(Z) = 5$$

c $Z = XY$

$$\mathrm{E}(Z) = \mathrm{E}(XY) = \mathrm{E}(X) \times \mathrm{E}(Y)$$

Notice that $5 = 3 \times \dfrac{5}{3}$.

Exercise 2C

1 Six unbiased coins are independently flipped and the number of heads obtained is recorded.

  a Let variable X denote the outcome of one such coin flip. Write the probability distribution table and find the expected value and variance of the variable X.

  b Explain why the whole experiment can be written as $Y = X + X + X + X + X + X$, where Y is a random variable representing the outcome of flipping six coins.

  c Find the expected value and variance of the random variable Y.

  d Use the empirical rule to determine all possible outcomes of Y, and comment on these outcomes.

2 An unbiased die is rolled four times, and the number of multiples of 3 obtained is recorded.

  a Let variable X denote the outcomes for one roll of the die. Write the probability distribution table and find the expected value and variance of the variable X.

  b Explain why the whole experiment can be written as $Y = X + X + X + X$, where Y is a random variable representing rolling the die four times.

  c Find the expected value and variance of the random variable Y.

3 Random variable X has the parameters $\mu = 3$ and $\sigma^2 = 4$.

  a Find the expected value and variance of $X + X + X$.

  b Find the expected value and variance of $3X$.

  c Show that $\mathrm{Var}(X + X + X + 3X) \neq \mathrm{Var}(6X)$.

4 Random variables X and Y have the parameters

$\mu_X = 2, \sigma_X^2 = 1$ and $\mu_Y = 5, \sigma_Y^2 = 3$.

Find the expected value and variance of

a $X + X + X + X + X$

b $Y + Y + Y$

c $X + X + X + X + X + Y + Y + Y$

5 Shariza is rolling a regular tetrahedral die with faces 1, 2, 3 and 4, and Shikma is independently rolling a biased die with faces 1, 1, 1, 2, 2 and 3 respectively. Let random variables X and Y represent the values of the outcomes obtained by Shariza and Shikma respectively.

a Draw the probability distribution tables and find the expected values of the variables X and Y.

Their friend is going to give them as many concert tickets as the product of the faces obtained. Let random variable Z represent the number of concert tickets Shariza and Shikma are going to earn in this game.

b Draw the probability distribution table and find the expected value of the variable Z and comment on your result.

6 Independent random variables X and Y have the expected values μ and $\mu + 2$ respectively. Given that $E(XY) = \mu^3$ find all possible non-zero values of μ.

A linear combination of independent normal random variables

In this section we are going to look at the particular case of a normal random variable. We start with a corollary that is a consequence of Theorem 2.

Corollary to Theorem 2

If we take two independent normal random variables
$X \sim N(\mu_1, \sigma_1^2)$ and $Y \sim N(\mu_2, \sigma_2^2)$, then the linear
combination $aX + bY$, $a, b \in \mathbb{R}$ is a normally distributed variable.

The expected value and variance of $aX + bY$ are
$\mu = a\mu_1 + b\mu_2$ and $\sigma^2 = a^2\sigma_1^2 + b^2\sigma_2^2$, so we can write
$aX + bY \sim N(a\mu_1 + b\mu_2, a^2\sigma_1^2 + b^2\sigma_2^2)$.

> Coefficients a and b cannot both be equal to 0 (we write this condition as $a^2 + b^2 \neq 0$).

Example 9

Given the normal random variables $X \sim N(3, 0.25)$ and $Y \sim N(4, 0.64)$, find the probabilities of the following:

a $X + Y > 8$

b $Y - X \le 0.5$

c $3X < 2Y$

a $X + Y \sim N(3 + 4, 0.25 + 0.64) = N(7, 0.89)$

$P(X + Y > 8) = 0.145$

Apply the corollary to thereom 2 and calculate the probabilities by using a GDC.

b $Y - X \sim N(4 - 3, 0.64 + 0.25) = N(1, 0.89)$

$P(Y - X \le 0.5) = 0.298$

c $3X < 2Y \Rightarrow 3X - 2Y < 0$

$3X - 2Y \sim N(3 \times 3 - 2 \times 4, \ 3^2 \times 0.25 + 2^2 \times 0.64)$

$= N(1, 4.97)$

$P(3X - 2Y < 0) = 0.327$

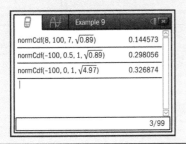

Notice that in part **a** the upper boundary is 100 and in parts **b** and **c** the lower boundaries are -100 which are sufficient values that represent $+\infty$ and $-\infty$ respectively.

Example 10

There are two classes at a school. One class studies the Mathematics Higher Level (HL) course, whilst the other class studies the Mathematics Standard Level (SL) course. The final grades of both classes are distributed normally, with the HL class having a mean of 5.2 and standard deviation of 1.37, whilst the SL class has a mean of 4.8 and standard deviation of 1.85. The HL class claims that they have obtained a better result than the SL class. Find the probability that this statement is true.

$X \sim N(5.2, 1.37^2)$

$Y \sim N(4.8, 1.85^2)$

$X > Y \Rightarrow X - Y > 0$

$X - Y \sim N(5.2 - 4.8, 1.37^2 + 1.85^2)$

$= N(0.4, 5.2294)$

$P(X - Y > 0) = 0.569$

Since the probability is greater than 0.5 we can conclude that the HL class have obtained a better result than the SL class.

Denote the HL and SL classes by the random variables X and Y respectively. Find the parameters of the difference of two random variables.

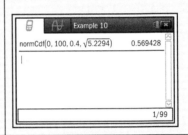

The corollary to Theorem 3 can now be generalized for n independent normal random variables.

Corollary to Theorem 3

If we take n independent normal random variables

$X_1 \sim N(\mu_1, \sigma_1^2),\ X_2 \sim N(\mu_2, \sigma_2^2), ...,\ X_n \sim N(\mu_n, \sigma_n^2),\ n \in \mathbb{Z}^+$

then the linear combination of $a_1 X_1 + a_2 X_2 + ... + a_n X_n = \displaystyle\sum_{k=1}^{n} a_k X_k,\ a_k \in \mathbb{R},\ \sum_{k=1}^{n} a_k^2 \neq 0$

is also a normal variable. The parameters are $\mu = \displaystyle\sum_{k-1}^{n} a_k \mu_k$ and $\sigma^2 = \displaystyle\sum_{k=1}^{n} a_k^2 \sigma_k^2$,

so we can write $\displaystyle\sum_{k=1}^{n} a_k X_k \sim N\left(\sum_{k=1}^{n} a_k \mu_k, \sum_{k=1}^{n} a_k^2 \sigma_k^2 \right)$.

The proof of this theorem can be done by mathematical induction. It is left as an exercise for the student.

Example 11

Given the following normal random variables $X \sim N(2, 2.25)$, $Y \sim N(4, 1.44)$, $Z \sim N(-3, 0.4)$ and $W \sim N(0.5, 0.64)$ calculate the following:

a $P(X - Y - Z > 0)$

b $P(2X - Z - 4W < 0)$

c $P(2X + Y > -3W - Z)$

d $P(X + 3Y < 2W - Y)$

---

a $X - Y - Z \sim N(2 - 4 + 3, 2.25 + 1.44 + 0.4)$

$\quad = N(1, 4.09)$

$\quad P(X - Y - Z > 0) = 0.690$

Find the mean and variance of the new variable and use a GDC to find the probability.

b $2X - Z - 4W$

$\quad \sim N(2 \times 2 - (-3) - 4 \times 0.5, 4 \times 2.25 + 0.4 + 16 \times 0.64)$

$\quad = N(5, 19.64)$

$\quad P(2X - Z - 4W < 0) = 0.130$

Find the mean and variance of the new variable and use a GDC to find the probability.

c $P(2X + Y > -3Z - W) = P(2X + Y + 3Z + W > 0)$

$\quad 2X + Y + 3Z + W$

$\quad \sim N(4 + 4 - 9 + 0.5, 4 \times 2.25 + 1.44 + 9 \times 0.4 + 0.64)$

$\quad = N(-0.5, 14.68)$

$\quad P(2X + Y > -3Z - W) = 0.448$

Rewrite the inequality so that we have one random variable and then find the mean and variance of the new variable. Use a GDC to find the probability.

d $P(X + 3Y < 2W - Y) = P(X + 3Y - 2W + Y < 0)$

$\quad X + 3Y - 2W + Y$

$\quad \sim N(2 + 12 - 1 + 4, 2.25 + 9 \times 1.44 + 4 \times 0.64 + 1.44)$

$\quad = N(17, 19.21)$

$\quad P(X + 3Y < 2W - Y) = 0.0000525$

Notice that we haven't simplified the expression because, for random variables, $3Y + Y \neq 4Y$ since the values of the variances of $3Y + Y$ and $4Y$ are different.

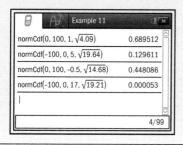

Example 12

Bill, Jill, and John are going to a steakhouse. There are three types of steak dishes they can order: X-large, Large, and Small. The weights of all the steaks are distributed normally and the parameters are given in the following table.

Steak	Average weight	Standard deviation
X-Large	430 g	30 g
Large	315 g	22 g
Small	150 g	8 g

Bill orders one X-Large steak, Jill orders one Large steak and John orders one Small steak. Find the probability that Bill will get a heavier steak than Jill's and John's steaks put together.

$X \sim N(430, 30^2)$, $L \sim N(315, 22^2)$, $S \sim N(150, 8^2)$

$P(X > L + S) = P(L + S - X < 0)$

$L + S - X \sim N(315 + 150 - 430, 22^2 + 8^2 + 30^2)$

$= N(35, 1448)$

$P(L + S - X < 0) = 0.179$

Denote the weights of X-large, Large, and Small steaks by random variables X, L, and S respectively.

Rewrite the inequality in a simpler form and construct a new variable. Find the parameters of the new variable.

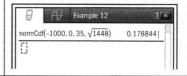

Example 13

Every day two wrestlers Kenji and Kazuyoshi go for a snack at a pizza place. Kenji always orders one Large pizza and Kazuyoshi orders one Jumbo pizza. Since Kazuyoshi needs more time to finish his pizza, Kenji gets hungry again and orders one small pizza.
The weights of all sizes of pizza are normally distributed. Large pizzas have a mean of 900 g and a variance of $25 \, g^2$, and Jumbo pizzas are 1.5 times the weight of Large pizzas. Small pizzas have a mean of 440 g and a variance of $10 g^2$.

a Find the mean and the variance of the weights of Jumbo pizzas.
b Find the probability that on a given day Kenji will eat more pizza by weight than Kazuyoshi.
c Find the probability that Kenji will eat more pizza by weight than Kazuyoshi during a three day period.

a Let L, J, and S represent the random variables of the weights of Large, Jumbo, and Small pizzas.

$J = 1.5 \times L \Rightarrow J \sim N(1.5 \times 900, 1.5^2 \times 25)$
$\quad = N(1350, 56.25)$

Use the formulas:

$E(aX) = aE(X)$

$Var(aX) = a^2Var(X)$

b $D = L + S - J \Rightarrow$

$D \sim N(900 + 440 - 1.5 \times 900, 25 + 10 + 1.5^2 \times 25)$

$\quad = N(-10, 91.25)$

$P(D > 0) = 0.148$

Let D represent the daily difference between the weight of pizza that Kenji and Kazuyoshi eat. Use a GDC to find the probability that D > 0.

c $X = D + D + D \Rightarrow$

$X \sim N(3 \times -10, 3 \times 91.25) = N(-30, 273.75)$

$P(X > 0) = 0.0349$

Let X represent the difference between the weights during a three day period.

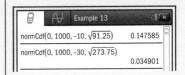

Exercise 2D

1 Given the following independent normal random variables
$X \sim N(0, 1)$, $Y \sim N(1, 0.16)$, $Z \sim N(-2, 0.25)$ and $W \sim N(3, 1.21)$
calculate the following:

a $P(Y - Z - W < 0)$

b $P(X + Y + Z + W > 0)$

c $P(3X + Y > Z + W)$

d $P(X - 3Z \leq 2Y + W)$

e $P(X - 4Z \leq 3X - 2Z)$

f $P(W - Y \leq 2Y + 3W)$

2 A market stall sells beetroot and sweet potato. The weights, in grams, are assumed to be normally distributed and the mean values and standard deviations are given in the table below.

	Mean	Standard deviation
Beetroot	240 g	20 g
Sweet potato	730 g	50 g

a Find the probability that the weight of a randomly chosen sweet potato is more than three times the weight of a randomly chosen beetroot.

b Find the probability that the weight of two randomly chosen sweet potatoes and four randomly chosen beetroots will exceed 2.5 kg.

3 Lillian and Veronica live in the same building and work at the same hospital. Lillian always takes a tram and a bus but Veronica walks and takes the tube. The times taken to travel to the hospital are assumed to be normally distributed and independent of each other. In the table below are the mean values and standard deviations of these times in minutes.

	Mean	Standard deviation
Lillian	35	5
Veronica	45	8

a Find the probability that on a given day Lillian will take less than 30 minutes to get to the hospital.

b Given that Veronica works five days a week, find the probability that within a week she will spend more than 4 hours travelling to work (not including her journey home again).

c Find the probability that, on a given day, four times the time taken by Lillian to get to the hospital is more than three times the time taken by Veronica to get to the hospital.

4 Dominic grows apples on his farm. It may be assumed that weights of the apples are normally distributed with a mean of 255 grams and a standard deviation of 12 grams.

a Find the probability that a randomly chosen apple weighs more than 250 grams.

b Four of these apples are selected at random. Find the probability that the total weight of the four apples is greater than 1 kilogram.

Dominic also grows plums. It may be assumed that weights of the plums are normally distributed with a mean of 60 grams and a standard deviation of 2 grams.

c Find the probability that the weight of a randomly chosen apple is more than four times the weight of a randomly chosen plum.

d Find the probability that the weight of a randomly chosen apple is more than the weight of four randomly chosen plums.

? Statistical sampling is very important for researchers in many disciplines; in order to draw conclusions for the entire population in a study they may have to select a small sample from the population. The main concern of these researchers is the representativeness of the sample: if it is not representative of the population then meaningful conclusions cannot be drawn, since the results obtained from the sample will be different from the results if the entire population were to be tested. However, when it comes to selecting the type of sampling technique, researchers must take in account other factors such as the duration of the study, ethical concerns and resources available.

2.2 Sampling distribution of the mean

In the core course we studied **population** and **samples**. We said that a population includes all the members of a defined group and a sample is just a representative selection or subset of the population. In order to be able to predict something about the population from a sample we need to have a **random sample**. A random sample is a sample where every element of the population has an equal opportunity to be selected to be part of the sample.

We are going to develop some ideas of sampling from a population. The study of normal probability distribution plays an important role in the theory of sampling, as it has a lot of practical applications.

For example, suppose a manufacturer of nails decides to check for irregular nails. The process of checking nails is not a simple one, since it would be very impractical to weigh all the nails individually, and there is a considerable variation in the weights of manufactured nails. The manufacturer needs to decide what weight range is acceptable.

To consider whether the entire population of nails are generally within the correct weight range, the manufacturer can take a sample. Taking a sample of a certain fixed size from a large population doesn't change the mean and the standard deviation of the population. The sample is taken independently, so we can consider the weights of nails in the sample to be independent normal random variables. However, what is the mean and standard deviation of weights of nails in the sample?

Let's remind ourselves what we have learned so far in the study of expectation algebra. If we have independent random variables $X_1, X_2, ..., X_n \sim N(\mu, \sigma^2)$, $n \in \mathbb{Z}^+$, then

$$E\left(\sum_{i=1}^{n} X_i\right) = n\mu \text{ and } Var\left(\sum_{i=1}^{n} X_i\right) = n\sigma^2$$

Now, however, we'll consider a variable $\bar{X}$ which is an independent normal random variable within a random sample of size k, where $k \leq n$. We'll look at the sample mean and its parameters.

i $\quad E(\bar{X}) = E\left(\dfrac{\sum_{i=1}^{n} X_i}{n}\right) = \dfrac{1}{n}E\left(\sum_{i=1}^{n} X_i\right) = \dfrac{1}{n} \times n\mu = \mu$

ii $\quad Var(\bar{X}) = Var\left(\dfrac{\sum_{i=1}^{n} X_i}{n}\right) = \dfrac{1}{n^2}Var\left(\sum_{i=1}^{n} X_i\right) = \dfrac{1}{n^2} \times n\sigma^2 = \dfrac{\sigma^2}{n}$

If the population variance is σ^2 and samples of size n are taken, then the distribution of the sample mean of these samples will have the same mean value μ as the population and the variance $\dfrac{\sigma^2}{n}$.

$$X \sim N(\mu, \sigma^2) \Rightarrow \bar{X} \sim N\left(\mu, \dfrac{\sigma^2}{n}\right)$$

The term **standard error** is used to describe the quantity $\dfrac{\sigma}{\sqrt{n}}$, which is the average distance from the mean that each random variable within the sample of size n will lie. By looking at the formula we notice that the larger the sample we take the smaller the standard error we obtain.

Example 14

The weight of trout in a fish farm may be assumed to be normally distributed with a mean of 340 g and a standard deviation of 30 g.

a Find the probability that a catch of 10 trout will have a mean weight per fish of more than 370 g.

b The hand net can hold up to 7 kg. Find the probability that we will be able to catch 20 trout in the hand net without breaking it.

- -

a $X \sim N(340, 30^2) \Rightarrow$

$$\overline{X} \sim N\left(340, \frac{30^2}{10}\right) = N\left(340, \left(\frac{30}{\sqrt{10}}\right)^2\right)$$

Use the formula for finding the mean and standard error of the sample distribution.

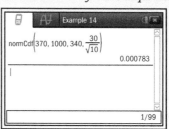

$$P(\overline{X} > 370) = 0.000783$$

b **Method I**

$$\overline{Y} \sim N\left(340, \left(\frac{30}{\sqrt{20}}\right)^2\right)$$

$$P\left(\overline{Y} < \frac{7000}{20}\right) = P(\overline{Y} < 350) = 0.932$$

Use the GDC to find the solution.

Method II

$$\overline{Y} \sim N\left(340, \left(\frac{30}{\sqrt{20}}\right)^2\right) \Rightarrow T = 20\overline{Y}$$

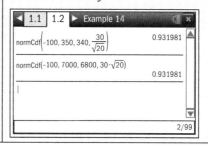

$$\Rightarrow T \sim N(6800, (30\sqrt{20})^2)$$

$$\Rightarrow P(T < 7000) = 0.932$$

Example 15

In a nail factory, the weights of all manufactured nails follow a normal distribution with a mean value of 8.2 g and a standard deviation of 0.02 g. Suppose that a control scale weighs 4 nails and the mean weight is recorded.

a Find the parameters of the sample mean distribution.

b By using the empirical rule, find the interval of weights that would be acceptable for all the samples of 4 nails.

- -

a $X \sim N(8.2, 0.02^2) \Rightarrow$

$$\overline{X} \sim N\left(8.2, \frac{0.02^2}{4}\right) = N\left(8.2, \left(\frac{0.02}{2}\right)^2\right)$$

Use the formula for finding the mean and standard error of the sample distribution.

b $8.2 \pm 3 \times 0.01 = 8.2 \pm 0.03$

We accept all the samples of 4 nails that have a mean weight between 8.17 g and 8.23 g.

Find the values which are 3 standard deviations above and below the mean.

The empirical rule in statistics states that almost all (99.7%) the observations fall within the range of three standard deviations from the mean value.

Example 16

Manufactured screws follow a normal distribution with a mean length of $3\,\text{cm}$ and a variance of $0.04\,\text{cm}^2$. What sample size should be taken to be 99% certain that the mean of the sample will lie within $0.1\,\text{cm}$ of the population mean length?

$X \sim N(3, 0.04^2) \Rightarrow$

$\overline{X} \sim N\left(3, \dfrac{0.04}{n}\right) = N\left(3, \left(\dfrac{0.2}{\sqrt{n}}\right)^2\right)$

$P(2.9 \leq \overline{X} \leq 3.1) = 0.99$

Use the formula for finding the mean and standard error of the sample distribution.

Method I

$P\left(\dfrac{2.9-3}{\dfrac{0.2}{\sqrt{n}}} \leq Z \leq \dfrac{3.1-3}{\dfrac{0.2}{\sqrt{n}}}\right) = 0.99$

Transform the variable into the standard normal variable by $Z = \dfrac{X-\mu}{\sigma}$.

Notice that the boundaries are symmetrical about the mean value so we can apply the property of the standard normal curve.

$P\left(Z \leq \dfrac{0.1\sqrt{n}}{0.2}\right) = 0.995$

$\dfrac{\sqrt{n}}{2} = \Phi^{-1}(0.995)$

$\sqrt{n} = 2 \times 2.57583 \Rightarrow \sqrt{n} = 5.15166$

$n = 5.15166^2 = 26.5396 \approx 27$

$P(-a < Z < a) = 1 - \alpha \Rightarrow P(Z < a) = 1 - \dfrac{\alpha}{2}$

Apply the inverse function.

Notice that the sample size must be a positive integer value therefore we round it to 27.

Method II

Find the solution by using the function features on your GDC.

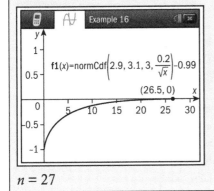

$n = 27$

Exercise 2E

1 Given that $X \sim N(\mu, \sigma^2)$ and the sample size n is taken from the population, find:

  a $P(1 \leq \bar{X} \leq 3)$ if $\mu = 1.5$, $\sigma^2 = 4$, $n = 10$

  b $P(|\bar{X} - 5| \leq 3)$ if $\mu = 5$, $\sigma^2 = 9$, $n = 7$

  c $P(|\bar{X}| \geq 0.8)$ if $\mu = -0.2$, $\sigma^2 = 8$, $n = 11$

2 Sleeping habits of students at a university are found to be distributed normally with the mean value of 5.5 hours and the standard deviation of 1.2 hours. 15 students from the university were selected at random. Find the probability that on average they sleep less than 5 hours.

3 A random sample of size 5 is taken from the normal distribution with the mean value 35 and the standard deviation 6.

  a Find $P(\bar{X} \geq 37)$.

  b Find the probability that the total sample value is more than 180.

4 The weights of buyers in a large shopping centre are distributed normally with the mean value of 62.5 kg and the standard deviation of 18.5 kg.

An elevator has a label that states a maximum load of 800 kg.

  a Find the probability that an average weight of a randomly selected group of 12 people will not exceed 70 kg.

  b Calculate the probability that the total weight of the randomly selected group of 12 people will not exceed the maximum load of the elevator.

2.3 The Central Limit Theorem

? Throughout history many mathematicians have worked on the Central Limit Theorem (CLT), one of the most important theorems in all of mathematics. The first mathematician to write about it was Abraham de Moivre. Almost a century later Pierre-Simon Laplace published it and extended de Moivre's work. During this period Cauchy, Dirichlet, Poisson, and Bessel made important contributions. Later on, George Pólya named the theorem "central" due to the importance of the CLT in probability theory. Along with Pólya, other mathematicians including von Mises, Lindeberg, and Lévy were working on the CLT. Finally, contributions by Chebyshev, Markov, and Lyapunov led to the first publication of the CLT in a general setting.

So far we have been taking samples only from normal populations. This time let's take an example that doesn't follow a normal distribution. Using a spreadsheet, we have simulated throwing a tetrahedral die four times, and then calculated the mean value of the four throws. We have performed 10 repetitions of a set of four throws (shown in column A below), and have calculated the mean of each set of 4 throws (shown in column B below). We have then repeated the process five more times (shown in columns D – Q).

	A	B	D	E	G	H	J	K	M	N	P	Q
15	2		1		2		1		3		4	
16	4	2,5	4	2,5	2	2	4	2,25	2	2,75	1	2,5
17	2		2		3		1		3		2	
18	2		4		2		2		2		3	
19	4		2		4		3		2		3	
20	4	3	2	2,5	1	2,5	1	1,75	2	2,25	3	2,75
21	2		4		3		2		3		2	
22	3		4		3		4		1		1	
23	3		2		4		3		3		1	
24	1	2,25	1	2,75	3	3,25	2	2,75	2	2,25	4	2
25	4		4		3		2		2		4	
26	4		3		1		4		1		4	
27	1		4		1		1		3		2	
28	3	3	1	3	3	2	2	2,25	2	2	4	3,5
29	1		3		1		4		2		3	
30	4		2		3		1		2		2	
31	1		1		1		2		1		2	
32	3	2,25	2	2	1	1,5	4	2,75	4	2,25	4	2,75
33	1		1		2		2		3		2	
34	4		3		2		2		1		3	
35	3		1		4		2		4		4	
36	1	2,25	4	2,25	2	2,5	3	2,25	1	2,25	4	3,25
37	2		4		4		4		4		2	
38	2		3		4		1		3		3	
39	1		3		3		1		2		2	
40	1	1,5	1	2,75	2	3,25	1	1,75	1	2,5	4	2,75
41	mean=	2,55	mean=	2,625	mean=	2,55	mean=	2,175	mean=	2,375	mean=	2,675
42	sd=	0,562731	sd=	0,428985	sd=	0,643342	sd=	0,373609	sd=	0,317324	sd=	0,457196

We know that a random variable X that describes the outcomes when throwing a tetrahedral die has the following probability distribution.

$X = x_i$	1	2	3	4
$P(X = x_i)$	$\dfrac{1}{4}$	$\dfrac{1}{4}$	$\dfrac{1}{4}$	$\dfrac{1}{4}$

We can see that $E(X) = \dfrac{1}{4}(1+2+3+4) = \dfrac{5}{2} = 2.5$ and

$$Var(X) = \frac{1}{4}(1+4+9+16) - \left(\frac{5}{2}\right)^2 = \frac{5}{4} \Rightarrow \sigma = \frac{\sqrt{5}}{2} = 1.12$$

In each repetition of four throws we notice that the sample mean values are in the range [2.175, 2.675] and the standard deviation values are

in the range [0.317324, 0.643342]. The mean values of the sample mean group themselves around the expected value of X, which is 2.5, while the standard deviation of X, which is 1.12, is not even in the range of the sample mean standard deviations.

We also notice that the standard error is $n = 4 \Rightarrow \dfrac{\sigma}{\sqrt{n}} = \dfrac{1.118033989}{2} = 0.559017$

and this value is in the obtained range [0.317324, 0.643342].

We are going to repeat the simulation for the sample size of $n = 10$ and $n = 50$.
We can see the results in the two spreadsheets below.

$n = 10$:

	A	B	D	E	G	H	J	K	M	N	P	Q
79	1		4		3		4		1		1	
80	3	2,6	1	2,4	3	3	3	2,8	3	2,7	3	2,8
81	4		1		2		3		3		2	
82	2		2		4		4		1		4	
83	4		3		2		1		2		3	
84	3		3		1		1		3		4	
85	4		2		2		1		1		4	
86	4		4		2		1		3		3	
87	1		3		1		3		2		2	
88	4		4		1		1		2		1	
89	1		2		1		2		2		2	
90	2	2,9	3	2,7	1	1,7	3	2	4	2,3	3	2,8
91	2		2		4		1		4		3	
92	4		2		3		2		2		4	
93	3		4		4		3		2		2	
94	1		1		1		4		4		4	
95	2		4		3		3		2		1	
96	2		4		3		2		4		1	
97	1		4		3		4		4		3	
98	1		1		2		1		3		1	
99	4		4		2		2		1		2	
100	2	2,2	1	2,7	2	2,7	2	2,4	2	2,8	2	2,3
101	mean=	2,53	mean=	2,56	mean=	2,46	mean=	2,54	mean=	2,62	mean=	2,44
102	s.d.=	0,275076	s.d.=	0,359629	s.d.=	0,392145	s.d.=	0,440202	s.d.=	0,285968	s.d.=	0,295146

In each repetition of ten throws, notice that the sample mean
lies within a range of [2.44, 2.62]. This is narrower than the range of the
sample mean for repetitions of four throws. The standard error is

$$n = 10 \Rightarrow \frac{\sigma}{\sqrt{n}} = \frac{1.118033989}{\sqrt{10}} = 0.35355, \text{ which is in the range of}$$

the standard deviations. $[0.275076, 0.440202]$

$n = 50$:

	A	B	D	E	G	H	J	K	M	N	P	Q
481	4		4		2		2		1		4	
482	1		4		1		3		2		2	
483	2		3		3		4		3		3	
484	2		3		2		3		3		3	
485	2		4		4		2		4		3	
486	1		1		1		3		3		3	
487	3		3		1		1		2		4	
488	1		1		2		3		1		3	
489	1		4		1		4		1		4	
490	1		3		4		3		3		2	
491	1		2		1		3		4		2	
492	2		4		3		3		3		1	
493	4		1		3		1		1		4	
494	2		2		2		2		3		3	
495	3		1		1		1		4		1	
496	3		2		4		2		2		3	
497	4		2		2		4		1		4	
498	2		4		1		2		1		2	
499	1		3		1		4		2		1	
500	1	2,28	2	2,4	3	2,34	1	2,5	1	2,42	4	2,46
501	mean=	2,514	mean=	2,552	mean=	2,442	mean=	2,508	mean=	2,444	mean=	2,516
502	s.d.=	0,224707	s.d.=	0,100311	s.d.=	0,114095	s.d.=	0,154402	s.d.=	0,101893	s.d.=	0,116924

In each repetition of 50 throws notice that the sample mean lies within an even narrower range [2.442, 2.552] than the range of the sample mean for repetitions of four throws or ten throws. The standard error for $n = 50$ is given by $n = 50 \Rightarrow \dfrac{\sigma}{\sqrt{n}} = \dfrac{1.118033989}{\sqrt{50}} = 0.15811$, which is in the range of the standard deviations [0.100311, 0.224707].

Our conclusion is that the larger the number of samples we take, the closer the parameters of the sample mean are getting to the population mean and the standard error.

X is a random variable with the parameters $E(X) = \mu$ and $Var(X) = \sigma^2$. Then as we take a large number of samples, the sample mean distribution approaches the normal distribution $\overline{X} \sim N\left(\mu, \dfrac{\sigma^2}{n}\right)$.

This is one of the most important and also most amazing facts in mathematics. Suppose we take any probability distribution (discrete or continuous) and take from it a sample of size n (n must be a large number). After taking a large number of those samples we look at the statistical characteristics of the means of all the samples taken. We should notice the following results:

- The mean of the samples is very close to the mean of the population that we took the samples from.
- The standard deviation of the samples is proportional to the standard deviation of the population divided by the factor that is close to the value of the square root of the sample size n, the standard error.
- With a large number of samples of size n, the distribution of the means of each sample can be better approximated by a normal distribution.

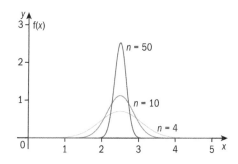

The simulation of the sampling distribution process and the Central Limit Theorem can be found at http://onlinestatbook.com/stat_sim/sampling_dist/ This website has both predefined uniform, normal, and skewed distributions, in addition to the capacity for students to create their own distributions, by graphing a probability distribution function and then applying the process of sampling distribution on it.

Theorem 5: The Central Limit Theorem

When sampling from **any population** X with the finite parameters μ and σ^2, the distribution of the sample mean $\overline{X}$ is approximately Normal if the sample size n is large enough. The mean of the distribution of the sample mean is equal to the population mean and the variance of the distribution of the sample mean is the variance of the parent population divided by the sample size,

$$\overline{X} \sim N\left(\mu, \dfrac{\sigma^2}{n}\right).$$

When we say "**any population**" we mean that we are not going to restrict ourselves to taking samples from a certain type of population such as a normal distributed population.

Example 17

From medical data collected in a school over a number of years, we can say that girls of age 15 who live in a large town have a mean height of 166 cm and a standard deviation of 6 cm. There are two sports teams made up of 5 girls and 8 girls within that age group.

a Find the probability that the mean height of the team with 5 girls is between 165 and 172 cm.

b Find the probability that the mean height of the team with 8 girls is between 165 and 172 cm.

c What can you conclude about those two teams?

---

a $\overline{X} \sim \mathrm{N}\left(166, \dfrac{6^2}{5}\right)$

$P(165 \leq \overline{X} \leq 172) = 0.633$

b $\overline{X} \sim \mathrm{N}\left(166, \dfrac{6^2}{8}\right)$

$P(165 \leq \overline{X} \leq 172) = 0.679$

Find the parameters of the sample mean.

Use a GDC to calculate the probability.

Find the parameters of the sample mean.

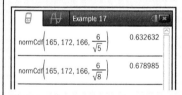

c With more members in a team, there is a higher probability that the team's mean height is closer to the mean height of the overall population of girls.

Notice that if we have more members of a team then we get a larger probability over the same interval of heights.

? To help assess a child's development, doctors use percentile growth charts to compare a child's height and weight to other children of the same age. Growth charts consist of a series of percentile curves that illustrate the distribution of selected body measurements in children. This valuable tool can help doctors determine whether a child is growing at an appropriate rate or whether there might be problems.

Example 18

A population with the parameters $\mu = 2$ and $\sigma^2 = 1$ is given. We take a sample of a size n from the population.

a Find the probability $P(1.5 \le \overline{X} \le 2.5)$ given that $n = 20$.

b Find the minimum sample size n such that $P(1.5 \le \overline{X} \le 2.5)$ is at least 0.9.

a $\overline{X} \sim N\left(2, \dfrac{1}{20}\right)$

$P(1.5 \le \overline{X} \le 2.5) = 0.975$

Find the parameters of the sample mean.

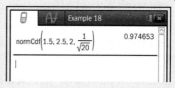

b $\overline{X} \sim N\left(2, \dfrac{1}{n}\right)$

$P(1.5 \le \overline{X} \le 2.5) = 0.9 \Rightarrow n = 11$

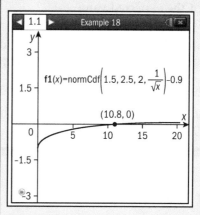

Round the value of n to the first larger integer.

Exercise 2F

1 Given the population parameters μ and σ^2 and the sample size n find:

a $P(1.5 \le \overline{X} \le 2.5)$ if $\mu = 2$, $\sigma^2 = 9$, $n = 30$

b $P(1.25 \le \overline{X} \le 1.35)$ if $\mu = 1.3$, $\sigma^2 = 0.04$, $n = 50$

c $P(\overline{X} \ge -0.48)$ if $\mu = -0.5$, $\sigma^2 = 1$, $n = 100$

d $P(\overline{X} < 397)$ if $\mu = 400$, $\sigma^2 = 234$, $n = 85$

e $P\left(|\overline{X}| < \dfrac{1}{2}\right)$ if $\mu = 1$, $\sigma^2 = 6$, $n = 40$

f $P\left(|\overline{X} - 2| \ge \dfrac{1}{5}\right)$ if $\mu = 1.9$, $\sigma^2 = 36$, $n = 120$

> **?** If the population distribution is symmetrical, then even with a smaller sample we are going to achieve a better approximation to a normal distribution. When the population distribution is not symmetrical then to achieve a good approximation to a normal distribution we need a larger sample. Usually we say that we need a sample size of at least $n = 30$ to achieve normality of the sampling distribution.

2 A real estate agent claims that the prices of studio apartments in a city have a mean value of €30,000 and standard deviation of €8,000. Given that 15 studio apartments were selected at random find the probability that the average price doesn't exceed €32,000.

3 The gestation period for dogs has a mean value of 63 days and a standard deviation of 2 days. Three female dogs are selected during this period. Find the probability that:

a the average gestation period of selected dogs will last longer than 64 days;

b the average gestation period of selected dogs will last less than 60 days;

c the average gestation period of selected dogs will last within 2 days of the expected value.

4 A population with the parameters $\mu = 6$ and $\sigma^2 = 4$ is given. We take a sample of size n from the population.

a Find the probability $P(5.4 \leq \overline{X} \leq 7.6)$ given that $n = 10$.

b Find the minimum sample size n such that $P(5 \leq \overline{X} \leq 7)$ is at least 0.95.

Review exercise

EXAM-STYLE QUESTIONS

1 Blanka and Svetlana are athletes specializing in the high jump. Blanka's jumps are normally distributed with a mean value of 2.02 m and a standard deviation of 4 cm. Svetlana's jumps are normally distributed with a mean value of 1.98 m and a standard deviation of 9 cm. Their jumps are independent. During a competition Blanka made five jumps and Svetlana made four jumps. Find the probability that the average height of Blanka's jumps was greater than the average height of Svetlana's jumps.

2 Let $X \sim \text{Po}(m)$ such that $\text{Var}(2X) = (\text{E}(X))^2 - 5$.

a Show that $m = 5$.

b Hence find $P(X \geq 6)$.

Another random variable Y, that is independent of X, has a Poisson distribution such that $\text{Var}(3Y) = 18$.

c Find $P(X + Y < 5)$.

Let random variable Z be such that $Z = 3X + 4Y$.

d Find the mean and variance of Z.

e State with a reason whether or not Z has a Poisson distribution.

3 A fish shop sells three types of fish: bass, bream, and cod. The weights of these fish may be assumed to be normally distributed. The mean values and standard deviations, in grams, are given in the table below.

Fish	Mean	Standard deviation
Bass	320	12
Bream	400	20
Cod	350	15

   a Find the probability that the weight of one bream exceeds 450 g.

   b Find the probability that the weight of two cod is less than 670 g.

   Nicholas buys one bass and one bream and Alex buys one cod.

   c Find the parameters of the total weight of fish Nicholas has bought.

   d Calculate the probability that the total weight of Nicholas' bass and bream is less than twice the weight of Alex's cod.

4 a A random variable X follows a binomial distribution and $\text{Var}(X) = 6$. Given that $n = 27$ find all the possible values of $\text{E}(3X - 7)$.

   b A random variable Z follows a Poisson distribution and

$$\left(\text{Var}(Z)\right)^2 = \text{E}(Z) + 12.$$

   Calculate $\text{Var}(5 + 2Z)$.

5 Prices of second-hand family cars advertised on a website have a mean value of €12,000 and a standard deviation of €3,200.

   a A car is selected at random. Find the probability that the price exceeds €13,000.

   b Given that 30 cars are selected at random from the website, find the probability that the average price of those 30 cars will be less that €11,500.

   c Vladimir is buying a second hand family car and he is willing to spend between €10,500 and €12,500. How many cars should he randomly select in order to have the probability of at least 85% that an average price falls within the desirable range?

6 a Let X be a random variable. By expanding the expression $\text{E}(X - \text{E}(X))^2$ show that $\text{E}(X^2) \geq (\text{E}(X))^2$.

   b Given two independent random variables such that $X \sim \text{Geo}(p)$ and $Y \sim \text{Geo}(q)$, where $p + q = 1$, show that $\text{Var}(X + Y) = \text{E}(X + Y)(\text{E}(X + Y) - 3)$.

Chapter 2 summary

Linear transformation of a single variable

Given that X is a random variable with finite parameters
and $a, b \in \mathbb{R}$ then

i) $E(aX + b) = aE(X) + b$

ii) $\text{Var}(aX + b) = a^2 \text{Var}(X)$

Linear transformation of two variables

Given that X and Y are two independent random variables with finite
parameters and $a, b \in \mathbb{R}$ then

i) $E(aX + bY) = aE(X) + bE(Y)$

ii) $\text{Var}(aX + bY) = a^2 \text{Var}(X) + b^2 \text{Var}(Y)$

Independent random variables

Given that independent random variables $X_1, X_2, X_3, \ldots, X_n, n \in \mathbb{Z}^+$ all have equal
expected value and equal variance, $E(X) = \mu$ and $\text{Var}(X) = \sigma^2$, then

$$E(X_1 + X_2 + \ldots + X_n) = n\mu \text{ and } \text{Var}(X_1 + X_2 + \ldots + X_n) = n\sigma^2$$

Independent normal random variables

If we take two independent normal random variables $X \sim N(\mu_1, \sigma_1^2)$ and $Y \sim N(\mu_2, \sigma_2^2)$
then the linear combination $aX + bY, \ a, b \in \mathbb{R}$ is going to be a normal variable also.
The parameters are $\mu = a\mu_1 + b\mu_2$ and $\sigma^2 = a^2\sigma_1^2 + b^2\sigma_2^2$, so we can write

$$aX + bY \sim N(a\mu_1 + b\mu_2, a^2\sigma_1^2 + b^2\sigma_2^2).$$

Sampling Distribution of the Mean

$E(\bar{X}) = \mu$ and $\text{Var}(\bar{X}) = \dfrac{\sigma^2}{n}$, Normal population sample

$$X \sim N(\mu, \sigma^2) \Rightarrow \bar{X} \sim N\left(\mu, \frac{\sigma^2}{n}\right)$$

The term **standard deviation of the mean**, $\dfrac{\sigma}{\sqrt{n}}$ is also known as the **standard error**.

The Central Limit Theorem

When sampling from **any population** X with the finite parameters μ and σ^2, the distribution of the sample mean $\bar{X}$ is approximately Normal if the sample size n is large enough. The mean of the distribution of the sample means is equal to the population mean and the variance of the distribution of the sample means is the variance of the parent population divided by the sample size, $\bar{X} \sim N\left(\mu, \dfrac{\sigma^2}{n}\right)$.

Exploring statistical analysis methods

CHAPTER OBJECTIVES:

7.3 Unbiased estimators and estimates; comparison of unbiased estimators based on variances;

$\bar{X}$ as unbiased estimator for μ. $\bar{X} = \sum_{i=1}^{n} \dfrac{X_i}{n}$ S^2 as unbiased estimator for σ^2.

$$S^2 = \sum_{i=1}^{n} \dfrac{\left(X_i - \bar{X}\right)}{n-1}$$

7.5 Confidence intervals for the mean of a normal population.

7.6 Null and alternative hypotheses H_0 and H_1. Significance level. Critical regions, critical values, p-values, one-tailed and two-tailed tests. Type I and II errors, including calculations of their probabilities. Testing hypotheses for the mean of a normal population.

Before you start

You should know how to:

1 Given the following table, find the mean and variance of the continuous variable X.

X	Frequency
$0 \le x < 10$	3
$10 \le x < 20$	12
$20 \le x < 30$	21
$30 \le x < 40$	18
$40 \le x < 50$	6

$\bar{x} = 27$, $\sigma^2 = 10.2956^2 = 106$

> Answers obtained by the GDC.

2 Given that $X \sim B(10, 0.3)$ find:

a $P(X = 5) = 0.103$

b $P(X \le 6) = 0.989$

c $P(1 \le X \le 3) = 0.621$

3 Given that $Y < Po(2)$ find:

a $P(Y = 0) = 0.135$

b $P(2 \le Y \le 5) = 0.577$

c $P(Y \ge 4) = 0.143$

Skills check:

1 Given the following table, find the mean and variance of the variable X.

X	Frequency
$0 \le x < 2$	22
$2 \le x < 4$	37
$4 \le x < 6$	46
$6 \le x < 8$	5

2 Given that $X \sim B(n, p)$ find:

a $P(X = 5)$ if $n = 5$ and $p = \dfrac{1}{2}$

b $P(3 \le X < 8)$ if $n = 10$ and $p = \dfrac{1}{5}$

3 Given that $Y < Po(m)$ find:

a $P(Y = 0)$ if $m = 0.4$

b $P(3 \le Y < 8)$ if $m = 7$

Biased information: how can we make sense of data?

Statistics are often presented as an attempt to add credibility to theories, proposals, ideologies, and arguments. You can see this in some television advertisements where many of the numbers put forward do not represent balanced statistical analysis (for examaple, a skin cream that promises 99% fewer wrinkles). They can be misleading and might entice you into making decisions that you find cause to regret later on. For these reasons, learning about statistics is a good step towards taking control of your life.

Statistical methods can be used to pull useful information out of huge amounts data. In our current digital age, this is extremely useful and possible to do systematically due to advancements in technology. Our task is to decide which statistics we want to use; calculators and computers can then be used to do the hard work for us! But what is statistics and how can we decide whether or not a particular statistic is balanced or relevant? Roughly, a statistic is a quantity calculated from a sample of data that tells us something about the properties of that sample. However, two different data samples from the same population can be very different. So, is there some way to accurately learn about an entire population from only a sampling of its values? Fortunately, there is a way, and it is called an *Estimator*.

An Estimator is a way of calculating a special type of statistic. This statistic reflects properties of both the data sample and also the entire population from which the sample was drawn. In this chapter we are going to study in detail how to obtain some estimators of parameters of a population, and analyse their quality in terms of how accurately they model the population's parameters.

3.1 Estimators and estimates

Suppose that you have to perform a statistical analysis for a company. You have lots of data, but where do you start? How can you use these samples of data to infer and draw conclusions about the entire population?

Let's say that you have a population that has a parameter, θ. Since you have no knowledge of the exact value of θ, we want to estimate this parameter by taking a random sample from the population. A random variable T describes the information that you obtain from that sample. The variable T is called an **estimator**. A specific **value** by which you approximate the parameter θ is called an **estimate**.

Look at the two normal distributions shown below. Consider two normal random variables, S and T, that are estimators for μ.

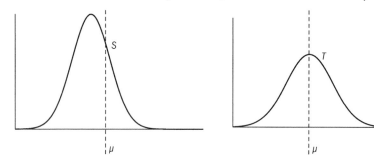

Notice that $E(S) \neq \mu$, whilst $E(T) = \mu$ since T has a symmetrical shape with respect to the mean value. In this case we say that T is an unbiased estimator for the mean value μ, and S is obviously biased and hence should not be used for estimation of the parameter.

In general a random variable T is called an unbiased estimator for the population parameter θ if $E(T) = \theta$.

Definition

An **estimator** of a population parameter (such as the mean, μ, or the variance, σ^2) **is a random variable** that depends on the sample data. The estimator provides an approximation to this unknown parameter. A specific **value** of that random variable is called an **estimate**.

Let us now consider two random variables, both unbiased estimators of the same parameter $E(T_1) = E(T_2) = \mu$, and consider which one might be a better estimator.

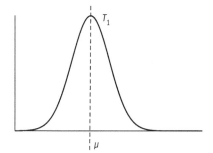 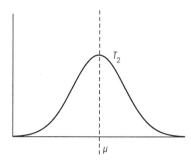

By looking at the graphs of both unbiased estimators T_1 and T_2, notice that T_1 has a smaller spread than T_2. Thus, the standard deviation (or variance) of T_1 is smaller than the standard deviation (or variance) of T_2. For a good estimation, it is essential to use a random variable with a small standard deviation.

Definition

Given two estimators T_1 and T_2 of the population we say that T_1 is a *more efficient estimator* than T_2 if $\mathrm{Var}(T_1) < \mathrm{Var}(T_2)$.

Example 1

Given two distributions of sample mean μ taken from the same normal population, show that the better estimator is the one with the larger sample.

$X_1, X_2, \ldots, X_n \sim N(\mu, \sigma^2),\ n \in \mathbb{Z}^+ \Rightarrow \overline{X} \sim N\left(\mu, \dfrac{\sigma^2}{n}\right)$ $Y_1, Y_2, \ldots, Y_m \sim N(\mu, \sigma^2),\ m \in \mathbb{Z}^+ \Rightarrow \overline{Y} \sim N\left(\mu, \dfrac{\sigma^2}{m}\right)$	*Take n observations and find the parameters of the sample mean. Take m observations from the same population and find the parameters of the sample mean.*
Assume that one is a better estimator: $\mathrm{Var}(\overline{X}) < \mathrm{Var}(\overline{Y}) \Rightarrow \dfrac{\sigma^2}{n} < \dfrac{\sigma^2}{m} \Rightarrow \dfrac{1}{n} < \dfrac{1}{m} \Rightarrow n > m$	*Use the definition of better estimator.*
Therefore the better estimator comes from a larger sample.	

Unbiased estimators for the mean and variance of a normal random variable

Two well-known statistics are the mean and variance. When creating statistical scenarios we will need to determine estimators for the mean and variance, i.e. we will need an unbiased estimator for the mean value of the set of normal random variables, and an unbiased estimator for the variance of the set of normal random variables.

Definition

For normal random variables $X_i \sim N(\mu, \sigma^2),\ 1 \leq i \leq n,$

1 $\overline{X}$ is an unbiased estimator for μ. $\overline{X} = \displaystyle\sum_{i=1}^{n} \dfrac{X_i}{n}$

2 S^2 is an unbiased estimator for σ^2. $S^2 = \displaystyle\sum_{i=1}^{n} \dfrac{(X_i - \overline{X})^2}{n-1}$

A well-defined definition?

We must show that the above definitions for unbiased estimators of μ and σ^2 are well defined. To do so, we must show the following:

1 $E(X) = \mu$

Since all of the samples are taken from the population $E(X_i) = \mu, i = 1, 2, ..., n$

$$E(\bar{X}) = E\left(\sum_{i=1}^{n} \frac{X_i}{n}\right) = \frac{1}{n} E\left(\sum_{i=1}^{n} X_i\right) = \frac{1}{n} \sum_{i=1}^{n} E(X_i) = \frac{1}{n} \times n\mu = \mu$$

2 $E(S^2) = \sigma^2$

By now we have found out that $E(\bar{X}) = \mu$ and $\text{Var}(\bar{X}) = \dfrac{\sigma^2}{n}$, but also

$E((X_i - \mu)^2) = \sigma^2, i = 1, 2, ..., n$ therefore we can calculate the following:

$$\sum_{i=1}^{n} E((X_i - \mu)^2) = n\sigma^2$$

$$\sum_{i=1}^{n} E((X_i - \mu)^2) = \sum_{i=1}^{n} E(((X_i - \bar{X}) + (\bar{X} - \mu))^2)$$

$$= \sum_{i=1}^{n} E((X_i - \bar{X})^2 + 2(X_i - \bar{X})(\bar{X} - \mu) + (\bar{X} - \mu)^2)$$

$$= \sum_{i=1}^{n} (E(X_i - \bar{X})^2 + 2E(X_i - \bar{X})E(\bar{X} - \mu) + E(\bar{X} - \mu)^2)$$

$$= \sum_{i=1}^{n} E(X_i - \bar{X})^2 + \sum_{i=1}^{n} 2E(\bar{X} - \mu)E(X_i - \bar{X}) + \sum_{i=1}^{n} E(\bar{X} - \mu)^2$$

$$= \sum_{i=1}^{n} E(X_i - \bar{X})^2 + 2E(\bar{X} - \mu)\underbrace{\sum_{i=1}^{n} E(X_i - \bar{X})}_{0} + \sum_{i=1}^{n} E(\bar{X} - \mu)^2$$

$$= \sum_{i=1}^{n} E(X_i - \bar{X})^2 + n\text{Var}(\bar{X}) = \sum_{i=1}^{n} E(X_i - \bar{X})^2 + n\frac{\sigma^2}{n}$$

$$= \sum_{i=1}^{n} E(X_i - \bar{X})^2 + \sigma^2$$

$$\Rightarrow n\sigma^2 = \sum_{i=1}^{n} E(X_i - \bar{X})^2 + \sigma^2$$

$$\Rightarrow \sum_{i=1}^{n} E(X_i - \bar{X})^2 = n\sigma^2 - \sigma^2 = (n-1)\sigma^2$$

S^2 is an unbiased estimator of the population variance because

$$E(S^2) = E\left(\sum_{i=1}^{n} \frac{(X_i - \bar{X})^2}{n-1}\right) = \frac{1}{n-1} E\left(\sum_{i=1}^{n} (X_i - \bar{X})^2\right) = \frac{1}{n-1}(n-1)\sigma^2 = \sigma^2$$

Thus, the above definitions for unbiased estimators of μ and σ^2 are well-defined. In examples 2 and 3, we will look at how to use these definitions.

Example 2

Show that the unbiased estimator of the variance can be found by using the formula:

$$s^2 = \frac{\sum_{i=1}^{n} x_i^2 - n(\overline{x})^2}{n-1}. \text{ Hence show that } s^2 = \frac{n}{n-1}\sigma^2.$$

$$s^2 = \frac{\sum_{i=1}^{n}(x_i - \overline{x})^2}{n-1} = \frac{\sum_{i=1}^{n}(x_i^2 - 2x_i\,x + (\overline{x})^2)}{n-1}$$	*Expand.*
$$= \frac{\sum_{i=1}^{n} x_i^2 - 2\overline{x}\sum_{i=1}^{n} x_i + \sum_{i=1}^{n}(\overline{x})^2}{n-1}$$	*Use the distributive property.*
$$= \frac{\sum_{i=1}^{n} x_i^2 - 2\overline{x} \times n\overline{x} + n(\overline{x})^2}{n-1} = \frac{\sum_{i=1}^{n} x_i^2 - n(\overline{x})^2}{n-1}$$	*Simplify the expression by using the definition of the mean.*
$$s^2 = \frac{\sum_{i=1}^{n} x_i^2 - n(\overline{x})^2}{n-1} = \frac{n}{n-1} \cdot \frac{\sum_{i=1}^{n} x_i^2 - n(\overline{x})^2}{n}$$	*Simplify by using the definition of variance.*
$$= \frac{n}{n-1}\left(\frac{\sum_{i=1}^{n} x_i^2}{n} - (\overline{x})^2\right) = \frac{n}{n-1}\sigma^2$$	

Example 3

Given the following set of data: 22, 24, 23, 22, 26, 26, 27, 25, 24, 24, 26, 25, 26, 27, 28 find:

a an unbiased estimation of the mean;

b an unbiased estimation of the variance.

Method I

a $\displaystyle \overline{x} = \frac{\sum x_i}{n} \Rightarrow \overline{x} = \frac{375}{15} = 25$

Unbiased estimate of the mean is the mean value of the sample itself.

b $\displaystyle s^2 = \frac{\sum_{i=1}^{n} x_i^2 - n(\overline{x})^2}{n-1} \Rightarrow s^2 = \frac{9421 - 15 \times 25^2}{14}$

Use the short formula for the unbiased estimate of the variance.

$\displaystyle = \frac{46}{14} = \frac{23}{7} = 3.29$

Method II

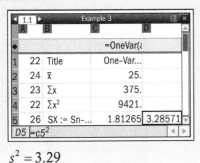

Use the GDC One-variable statistics calculation. In order to find the variance, square the unbiased estimate of the population standard deviation s.

$s^2 = 3.29$

Exercise 3A

1 Calculate the unbiased estimate of the mean and variance for the following sets of data:

    a {2, 4, 6, 8, 10, 12, 14, 16, 18, 20}

    b {21, 24, 36, 28, 30, 22, 25, 26, 38, 32, 34, 29, 37, 33, 31, 30}

    c {1, 4, 7, 10, ..., 133}

2 The distribution of broken eggs in 40 boxes (where each box contains 10 eggs) is given in the following table:

x_i	0	1	2	3
f_i	22	12	4	2

    Find unbiased estimates of the mean and standard deviation of the broken eggs.

3 The following table displays the times, T (minutes), taken by a group of 70 students to travel to school.

T	$0 \le t < 10$	$10 \le t < 20$	$20 \le t < 30$	$30 \le t < 60$	$60 \le t < 120$
Frequency	6	13	26	17	8

    Find:

    a an unbiased estimate for the mean of T

    b an unbiased estimate for the standard deviation of T.

3.2 Confidence intervals for the mean

Let's analyze a real-life scenario:

A large poultry farm raises over 1,000,000 broilers (young chickens) per year. After six weeks, broilers are ready to be released from the farm. It is difficult to measure the weight of every single released broiler, so a certain-size sample of six-week old broilers are weighed. The mean weight of a broiler from this sample is found to be 2.10 kg. Using the sample mean, we wish to estimate the mean value of the whole population of six-week-old broilers.

However, how confident can we be that the population mean lies within certain limits (in kilograms) of the sample mean value of 2.10 kg? Can we quantify such a confidence that the population mean lies within the prescribed interval?

Our ability to answer such a question depends on the information available to us, as the calculations to find such a confidence interval depend on how much we know about the population in question. Let's assume that the given data satisfies the conditions of the Central Limit Theorem.

> If data satisfies the conditions of the Central Limit Theorem, we are able to approximate the distribution of the data by a normal distribution. Using this approximation, we are able to calculate probabilities of events related to the real-life problem we're studying.

There are two types of situation that you may be faced with when calculating confidence intervals for the mean of a population:

Case I: The population standard deviation is known;

Case II: The population standard deviation is not known, and we have to estimate it from the sample.

Case I: Confidence interval for μ when σ is known

First we are going to investigate the case when population standard deviation is known.

If a random variable X follows a normal distribution such that $X \sim \mathrm{N}(\mu, \sigma^2)$ then, for any value $n, n \in \mathbb{Z}^+$, the sample mean is also normal and $\overline{X} \sim \mathrm{N}\left(\mu, \dfrac{\sigma^2}{n}\right)$. Let's take the **confidence value** to be $1 - \alpha$, where α is called the **level of significance**. This means that we can calculate an interval $[a, b]$ in which we are $(1 - \alpha)\%$ sure that the sample mean lies. The most commonly used values for the level of significance are $\alpha = 0.1, 0.05, 0.01$.

Sir Ronald Aylmer Fisher (1890–1962) was an English statistician, evolutionary biologist, geneticist, and eugenicist. He is considered by many to be "The father of statistics". He made immense contributions to the development of analysis of variance, experimental design, and likelihood based methods. In the early 20th century he developed statistical hypothesis testing and established the standard of significance (α) level to be 5%. The other most commonly used significant levels, 1% and 10%, were developed in relation to that particular field of study. Fisher was also one of the founders of population genetics. He contributed to biostatistics and biometrics, and was one of the most important biologists since Darwin.

We need to find the boundaries a and b of the confidence interval $[a, b]$ such that $\mathrm{P}(a \le \mu \le b) = 1 - \alpha$.

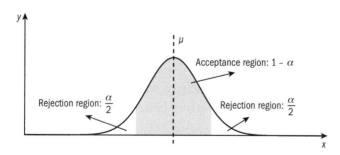

To find the boundaries we need to convert it to the standard normal distribution $z = \dfrac{\overline{x} - \mu}{\dfrac{\sigma}{\sqrt{n}}}$.

$$\Phi^{-1}\left(1 - \frac{\alpha}{2}\right) = z_{\frac{\alpha}{2}} \Rightarrow P\left(-z_{\frac{\alpha}{2}} \le z \le z_{\frac{\alpha}{2}}\right) = 1 - \alpha$$

$$\Rightarrow P\left(-z_{\frac{\alpha}{2}} \le \frac{\overline{x} - \mu}{\dfrac{\sigma}{\sqrt{n}}} \le z_{\frac{\alpha}{2}}\right) = 1 - \alpha$$

$$\Rightarrow P\left(-\frac{\sigma}{\sqrt{n}} \times z_{\frac{\alpha}{2}} \le \overline{x} - \mu \le \frac{\sigma}{\sqrt{n}} \times z_{\frac{\alpha}{2}}\right) = 1 - \alpha$$

$$\Rightarrow P\left(\overline{x} - \frac{\sigma}{\sqrt{n}} \times z_{\frac{\alpha}{2}} \le \mu \le \overline{x} + \frac{\sigma}{\sqrt{n}} \times z_{\frac{\alpha}{2}}\right) = 1 - \alpha$$

So in this case the confidence interval for the mean value of the population is given by the following formula:

$$\left[\overline{x} - \frac{\sigma}{\sqrt{n}} \times z_{\frac{\alpha}{2}}, \ \overline{x} + \frac{\sigma}{\sqrt{n}} \times z_{\frac{\alpha}{2}}\right]$$

The values of $z_{\frac{\alpha}{2}}$ are standard values for each α. The most common cases of confidence intervals are listed in the table below:

α	$z_{\frac{\alpha}{2}}$	$\left[\overline{x} - \dfrac{\sigma}{\sqrt{n}} \times z_{\frac{\alpha}{2}}, \ \overline{x} + \dfrac{\sigma}{\sqrt{n}} \times z_{\frac{\alpha}{2}}\right]$
0.1	$\Phi^{-1}(0.95) = 1.645$	$\left[\overline{x} - 1.645\dfrac{\sigma}{\sqrt{n}}, \ \overline{x} + 1.645\dfrac{\sigma}{\sqrt{n}}\right]$
0.05	$\Phi^{-1}(0.975) = 1.960$	$\left[\overline{x} - 1.960\dfrac{\sigma}{\sqrt{n}}, \ \overline{x} + 1.960\dfrac{\sigma}{\sqrt{n}}\right]$
0.01	$\Phi^{-1}(0.995) = 2.576$	$\left[\overline{x} - 2.576\dfrac{\sigma}{\sqrt{n}}, \ \overline{x} + 2.576\dfrac{\sigma}{\sqrt{n}}\right]$

Even though we can use a GDC to obtain confidence intervals easily, we will illustrate the use of both methods for finding confidence intervals to better understand their meaning.

Example 4

In a certain country a random sample of 100 men is taken. The mean height is found to be 183.6 cm, with a standard deviation of 5 cm. Find the 90% confidence interval for the mean height of the male population in the country, giving your answer correct to two decimal places. Interpret your answer.

Method I

$\bar{x} = 183.6$, $\sigma = \dfrac{5}{\sqrt{100}} = 0.5$

$\alpha = 0.1$

$\Rightarrow [183.6 - 1.645 \times 0.5, 183.6 + 1.645 \times 0.5]$

$\Rightarrow [182.7775, 184.4225]$

$\Rightarrow [182.8, 184.4]$

Identify the mean value of the sample and find the standard error.

Identify the significance level and apply the formula $\left[\bar{x} - 1.645 \dfrac{\sigma}{\sqrt{n}}, \bar{x} + 1.645 \dfrac{\sigma}{\sqrt{n}} \right]$.

Method II

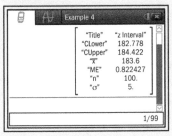

$\Rightarrow [182.8, 184.4]$

We are 90% confident that the population mean will lie within this interval.

Use a GDC to find the confidence interval. In the Statistics menu, choose Confidence Intervals, select z Interval, and then set the Data input method as 'Stats'.

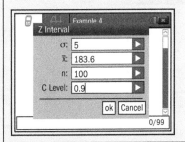

When entering the statistics into your GDC, you must enter the population standard deviation and not the sample standard deviation. The GDC uses the number of statistics (*n*) to calculate the sample's standard deviation itself.

Thus far, we have calculated confidence intervals for the mean only from statistics which represent the data (i.e. we are given the mean, standard deviation, and size of sample, from which we calculate the confidence interval). However, sometimes you may be asked to calculate confidence intervals from raw data. In this case you must calculate the statistics first (i.e. mean, standard deviation, etc.) and then proceed to calculate the confidence interval. We will look at doing this in Example 5.

Example 5

After a rainy night, 12 worms have surfaced on sand. Their lengths, measured in cm, were as follows: 12.0, 11.1, 10.5, 10.8, 12.1, 10.4, 10.9, 12.2, 10.9, 11.9, 11.2, and 11.6. Assuming that this sample came from a population that follows a normal distribution with variance 4, calculate the 95% confidence interval of the mean length of the worm population within the sand, and interpret your answer.

Method I

$$\bar{x} = \frac{\sum x_i}{12} = 11.3, \ \sigma = \frac{2}{\sqrt{12}}$$

$\alpha = 0.05$

$$\Rightarrow \left[11.3 - 1.960 \times \frac{2}{\sqrt{12}}, \ 11.3 + 1.960 \times \frac{2}{\sqrt{12}} \right]$$

$\Rightarrow [10.17, 12.43]$

Identify the mean value of the sample and find the standard error.

Identify the significance level and apply the formula $\left[\bar{x} - 1.960\frac{\sigma}{\sqrt{n}}, \ \bar{x} + 1.960\frac{\sigma}{\sqrt{n}} \right].$

Method II

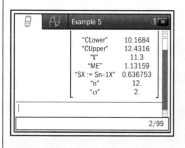

$\Rightarrow [10.17, 12.43]$

We are 95% confident that the population mean will lie within this interval.

Use a GDC to find the confidence interval. First, enter the lengths of the worms as a data list. Then, in the Statistics menu, choose Confidence Intervals, select z Interval, and then set the Data input method as 'Data'.

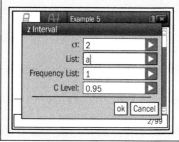

In examples 4 and 5, if we conduct the experiments a large number of times, approximately 90% and 95% of the intervals respectively will contain the population mean.

When planning a statistical survey, you may actually need to calculate the sample size you must choose in order to obtain a specific confidence interval at a certain distance from the mean. The following example shows how to calculate the required sample size.

Example 6

Wooden beams are manufactured for interior construction and the length of beams has a standard deviation of 2 cm. Find how many beams must be sampled so that we would be 99% confident that the sample mean doesn't differ from the beams' mean by more than 1 cm.

$\lvert \bar{x} - \mu \rvert \leq 1 \Rightarrow \lvert \mu - \bar{x} \rvert \leq 1$	*Rewrite as an inequality.*
$\Rightarrow -1 \leq \mu - \bar{x} \leq 1$	*Use the 99% confidence interval*
$\Rightarrow \bar{x} - 1 \leq \mu \leq \bar{x} + 1$	$\left[\bar{x} - 2.576\dfrac{\sigma}{\sqrt{n}}, \bar{x} + 2.576\dfrac{\sigma}{\sqrt{n}} \right]$
$\sigma = 2 \Rightarrow 2.576 \times \dfrac{2}{\sqrt{n}} = 1$	
$\Rightarrow \sqrt{n} = 4.514 \Rightarrow n = 20.376$	*Use the population standard deviation and find the sample size.*
We should take at least 21 beams.	*The sample size must be a positive integer.*

Exercise 3B

1 Given the values of $\bar{x}$, σ, n, and α, find a $(1 - \alpha)$% confidence interval for the mean if:

   a $\bar{x} = 5$, $\sigma = 1$, $n = 10$ and $\alpha = 0.01$

   b $\bar{x} = -11$, $\sigma = 4.3$, $n = 22$ and $\alpha = 0.05$

   c $\bar{x} = 2854$, $\sigma = 327$, $n = 230$ and $\alpha = 0.1$

2 Given the sample A from a population with standard deviation σ, find the $(1 - \alpha)$% confidence interval for the mean if:

   a $A = \{1, 2, 3, 4, 5, 6, 7, 8, 9\}$, $\sigma = 2$ and $\alpha = 0.1$

   b $A = \{0.1, 0.12, 0.15, 0.18, 0.13, 0.12, 0.09, 0.11, 0.13, 0.21, 0.15\}$, $\sigma = 0.04$ and $\alpha = 0.01$

   c $A = \{321, 325, 330, 324, 325, 326, 317, 318, 329, 310, 314, 318, 327, 322, 328\}$, $\sigma = 4.5$ and $\alpha = 0.05$

3 A sample with n elements is taken from a normal population that has a standard deviation of 5.5. Find the value of n so that we would be 95% confident that the sample mean doesn't differ from the population mean by more than 2.

4 Nayla buys 8 clementines and weighs them. Their weights, shown in grams (g), are: 70, 75, 77, 71, 68, 80, 85, 72. We may assume that these clementines came from a normal population with standard deviation of 3.5 g. Determine a 95% confidence interval for the mean weight of clementines.

5 The measurement of n independent random measurements may be assumed to follow a normal distribution with the mean value μ and the variance $\sigma^2 = 9$. Given that the 90% confidence interval for the mean is found to be $[14.2, 17.4]$ find:

a the mean value of the sample

b the sample size n.

In order to better understand the influence of the sample size on the length of a confidence interval, we are going to look at the following investigation.

Investigation

Let's consider samples of different sizes, all of which have the same mean value $\bar{x} = 100$. The samples come from a population with standard deviation $\sigma^2 = 64$.

a Find a 95% confidence interval for the mean for the following values of sample size n given that:

i $n = 10$

ii $n = 25$

iii $n = 50$

iv $n = 150$

b What can you conclude about the relationship between the lengths of intervals and sample sizes? Justify your answer.

Case II: Confidence interval for μ when σ is unknown

In real-life situations, populations constantly grow or decay, so it is very unlikely that we will know the parameters of the observed population. Even if we assume that it is normally distributed, there is hardly any situation when we are certain about population parameters. Thus, we often need to estimate them.

Investigation 1 illustrated that larger samples give a greater degree of certainty to our estimates. However, it is often difficult to take a large sample for practical reasons. We would, however, like to be as accurate as possible in our conclusions, when working with small or large samples.

Before we estimate the population standard deviation σ, we must first define a concept that we will use to make our estimate:

When σ is unknown we must first estimate it from available data. If we have a sample of size n, since we need to calculate an estimation of σ, we say that we have $v = n - 1$ degrees of freedom.

> A degree of freedom, v, is the number of data that can vary, without changing the population parameter we are estimating. For example, if we have a sample of size n and we have calculated an estimation of the standard deviation of the population, then $n - 1$ data can vary whilst the last one must be prescribed by the previous $n - 1$ data pieces in order to obtain the standard deviation we require.

To estimate the population standard deviation σ we use an unbiased estimate of the population standard deviation $s = \sigma\sqrt{\dfrac{n-1}{n}}$.

The Central Limit Theorem suggests that we can approximate the distribution of a large sample by a normal distribution, but the question still remains: How are smaller samples distributed?

> **?** William Sealy Gosset (1876 – 1937) worked as a statistician in the quality control department of a brewery. In 1908, he published a paper titled 't-test' under the pseudonym 'Student' because his employer strictly forbade employees from publishing any scientific papers. As a result, his name is not as well-known as the important statistical results he formalized. The t-test was initially used to ensure quality control of small samples of brewed beer. Gosset discovered the form of the t-distribution by a combination of mathematical and empirical work with random numbers; an early application of the Monte-Carlo method. At the end of 1935, Gosset left Ireland to take charge of the new Guinness brewery in London. Despite the hard work involved in this venture, he continued to publish statistical papers.

To approximate the distribution of smaller samples, we use t-distribution (commonly called *Student's t-distribution* after Gosset) when sample size doesn't exceed 30.

For a t-distributed random variable T, we approximate the standard deviation σ by the estimate of the standard deviation, s:

$$Z = \frac{\overline{X} - \mu}{\dfrac{\sigma}{\sqrt{n}}} \Rightarrow T = \frac{\overline{X} - \mu}{\dfrac{s}{\sqrt{n}}}$$

where the variable T follows a t-distribution with $n-1$ degrees of freedom and we write $T \sim t(n-1)$.

Student's t-distribution graphs are very similar to the standard normal distribution graph. They are bell-shaped curves; symmetrical about the y-axis and achieving maximum value when the curve crosses the y-axis.

Let's use a GDC to look at the probability density function of the standard normal distribution and some t-distributions with different degrees of freedom.

The graphs show that this distribution is aligned with the standard normal distribution for the larger degrees of freedom (i.e. for larger samples), but for smaller samples the *t*-distribution better approximates the data since the tail areas are a bit larger and not as insignificant as under the normal distribution.

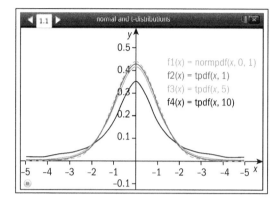

Since the standard normal distribution and the *t*-distribution curves are similar in shape, we can adapt the formula for the confidence interval of the mean of a normal distribution to find a formula for the confidence interval of the mean of a *t*-distribution.

Before the introduction of GDC technology, *z*-distribution and *t*-distribution tables were used to find critical values. Here is an example of a *t*-distribution table.

$p = P(X \leq t)$

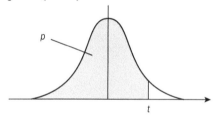

P	0.9	0.95	0.975	0.99	0.995	0.9995
$v = 1$	3.078	6.314	12.706	31.821	63.657	636.619
2	1.886	2.920	4.303	6.965	9.925	31.599
3	1.638	2.353	3.182	4.541	5.841	12.924
4	1.533	2.132	2.776	3.747	4.604	8.610
5	1.476	2.015	2.571	3.365	4.032	6.869

Nowadays calculators have a built-in "invt" feature, i.e. a function with two variables that calculates the critical *t* value; $t_c = invt\left(1 - \dfrac{\alpha}{2}, v\right)$.

So the confidence interval can be calculated by using the formula:

$$\left[\bar{x} - \frac{s}{\sqrt{n}} \times t_c, \ \bar{x} + \frac{s}{\sqrt{n}} \times t_c \right]$$

Example 7

In a sample of five $100\,g$ chocolates the mean value of the energy level was found to be $\bar{x} = 2540\,kJ$, whilst the estimate for the population standard deviation was $s = 120\,kJ$. Calculate the 99% confidence interval for the mean value of the energy level of $100\,g$ of chocolates.

Method I

$n = 5 \Rightarrow v = 4 \Rightarrow t_c = 4.604$

$2540 \pm \dfrac{120}{\sqrt{5}} \times 4.604$

$\Rightarrow \mu \in [2293, 2787]\,kJ$

$v = n - 1$ and use a GDC to find the critical
t-value.

Use the formula $\bar{x} \pm \dfrac{s}{\sqrt{n}} \times t_c$

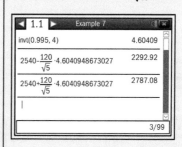

Method II

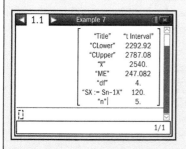

$\Rightarrow \mu \in [2293, 2787]\,kJ$, given correctly to the nearest kJ.

Use the GDC t-interval stats input method and input the given statistics.

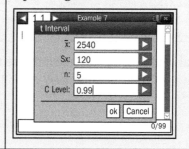

Let's try to see how to solve the problem of the mean weight of broilers.

Example 8

A large poultry farm grows broilers. A sample of 20 broilers is taken. The mean weight of the sample was found to be 2.10 kg. It is assumed that broilers come from a normal population with the unbiased estimate of the standard deviation of 0.3 kg. Calculate the 90% interval for the mean weight of a broiler.

Method I

$n = 20 \Rightarrow v = 19 \Rightarrow t_c = 1.729$

$2.10 \pm \dfrac{0.3}{\sqrt{20}} \times 1.729$

$\Rightarrow \mu \in [1.98, 2.22]$ kg

$v = n - 1$ and then use a calculator to find the t-value.

Use the formula $\bar{x} \pm \dfrac{s}{\sqrt{n}} \times t_c$

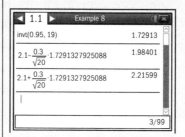

Method II

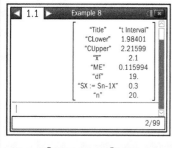

$\Rightarrow \mu \in [1.98, 2.22]$ kg

Use the GDC t-interval stats input method and input the given statistics.

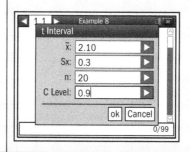

When the actual data is given, we need to use the list and statistic data features on the GDC, as shown in the following example.

Example 9

In a box of six eggs, the weights (measured in grams) of each egg were: 62, 63, 65, 62, 66, 66. Calculate the 95% confidence interval for the mean weight of an egg.

Method I

$$\bar{x} = \frac{\sum x_i}{n} \Rightarrow \bar{x} = \frac{384}{6} = 64$$

Find the mean value of the sample.

$$s = \sqrt{\frac{\sum x_i^2 - n(\bar{x})^2}{n-1}} \Rightarrow s = \sqrt{\frac{24\,594 - 24\,576}{5}} = 1.897$$

$$n = 6 \Rightarrow v = 5 \Rightarrow t = 2.571$$

$$64 \pm \frac{1.897}{\sqrt{6}} \times 2.571$$

$$\Rightarrow \mu \in [62, 66]\,\text{g}$$

Find the unbiased estimate of the population standard deviation. $v = n - 1$ and then we look at the tables or use a calculator to find the t-value.

Use the formula $\bar{x} \pm \dfrac{s}{\sqrt{n}} \times t_c$

Method II

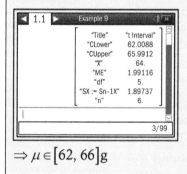

Use the GDC t-interval data input method where data is stored in a list.

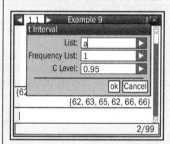

$$\Rightarrow \mu \in [62, 66]\,\text{g}$$

Example 10

A random sample of ten independent observations are taken from a normal population. The sample gave the results $\sum\limits_{i=1}^{10} x_i = 527$ and $\sum\limits_{i=1}^{10} x_i^2 = 28\,157$.

Find a 90% confidence interval for the population mean.

$$\bar{x} = \frac{\sum x_i}{n} \Rightarrow \bar{x} = \frac{527}{10} = 52.7$$

$$s = \sqrt{\frac{\sum x_i^2 - n(\bar{x})^2}{n-1}}$$

$$\Rightarrow s = \sqrt{\frac{28\,157 - 10 \times 52.7^2}{9}} = 6.5328$$

First we must find the unbiased estimate of the mean of the sample and then the unbiased estimate of the population standard deviation.

Now, use the GDC t-interval stats input method and input the given statistics.

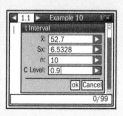

$$\mu \in [48.9, 56.5]$$

Exercise 3C

1 Given the values of $\bar{x}$, s, n, and α, find $(1 - \alpha)\%$ confidence interval for the mean if:

  a $\bar{x} = 15$, $s = 1.2$, $n = 15$ and $\alpha = 0.1$

  b $\bar{x} = -23$, $s = 5.8$, $n = 32$ and $\alpha = 0.01$

  c $\bar{x} = 3478$, $s = 429$, $n = 310$ and $\alpha = 0.05$

2 Given the set A of data find the $(1 - \alpha)\%$ confidence interval for the mean if:

  a $A = \{1, 2, 3, 4, 5, 6, 7, 8, 9\}$ and $\alpha = 0.01$

  b $A = \{0.1, 0.12, 0.15, 0.18, 0.13, 0.12, 0.09, 0.11, 0.13, 0.21, 0.15\}$ and $\alpha = 0.05$

  c $A = \{321, 325, 330, 324, 325, 326, 317, 318, 329, 310, 314, 318, 327, 322, 328\}$ and $\alpha = 0.1$

3 Nadim buys 12 mandarins and weighs them with the following results given in grams (g): 66, 70, 75, 84, 90, 77, 71, 68, 80, 85, 72, 63. We may assume that the mandarins came from a normal population. Determine a 99% confidence interval for the mean weight of the mandarins.

4 Eight independent random measurements are taken from a normal population with the unbiased estimate of population variance $s^2 = 2.25$. Given that the confidence interval for the mean is found to be $[12.345, 14.355]$ find:

  a the mean value of the sample

  b the confidence level of the interval.

5 A box with 15 cod is bought at a fish market. The mean weight of a cod at the market is 536 g. Weights of cod bought at the market are normally distributed and the unbiased estimate of the standard deviation is 95 g.

  a Find a 95% confidence interval for the mean weight of cod bought at the market.

  b Find a 99% confidence interval for the mean weight of cod bought at the market.

  c Comment on the significance level and the length of the confidence interval.

Confidence interval for matched pairs

Matched pairs are different samples taken from the same population. In our study, we want to know whether or not two matched pairs differ. For example, suppose the same product is manufactured in two different factories. We might take two samples of product, one from each factory, and compare the means of the two samples in order to determine whether the different samples satisfy the same standards of production.

Matched pairs can also be used to compare a variable *against itself* when measured in two different circumstances. An example might be in measuring the iron content of a group of people who suffer from a blood condition. We would measure the same group both before and after the medical treatment, and these two sets of measurements on the same group would form our matched pair. This sort of analysis is used to determine whether or not a new drug is effective.

Alternatively, we might use matched pairs analysis to study a group of people who follow a certain diet in order to lose weight. We might measure their weights before and after the diet to see whether they have achieved their goal, and the two sets of results *from the same group* would form our matched pair.

> **?** Matched pairs are used a lot in biological experiments. To determine the effect of a drug on a population, we might monitor two groups: an experimental group and a control group. We match and compare all the elements of one group with the elements of the other group based on certain characteristics (such as gender, age, weight, etc.). We would then compare the impact the drug has had on the experimental group, compared with the control group. This is done to minimize the effect of other influences on the experimental group, which, without a control group, we might ascribe to the effects of the drug.

Example 11

A group of patients at a hospital suffer from chronic iron deficiency. They are treated with two new drugs: drug A and drug B. First, they were treated with drug A and, after a few days when the influence of drug A had worn off, they were treated with drug B. Serum iron tests were conducted and the results (in $\mu g/dl$) are shown in the following table:

Patient	a	b	c	d	e	f	g	h
Drug A	60	58	47	80	35	55	53	40
Drug B	63	55	42	76	29	61	51	41

a Find the differences between the results obtained by drug A and drug B.
b Calculate the 90% confidence interval of the mean of differences in part **a**.

a

$d_i = A - B$	-3	3	5	4	6	-6	2	-1

Subtract the results of drug B from those of drug A.

b Method I

$$\bar{d} = \frac{\sum d_i}{n} \Rightarrow \bar{d} = \frac{10}{8} = 1.25$$

Calculate the mean of differences.

$$s_d = \sqrt{\frac{\sum d_i^2 - n(\bar{d})^2}{n-1}} \Rightarrow s_d = \sqrt{\frac{136 - 12.5}{7}} = 4.200$$

Calculate the unbiased estimate of the standard deviation.

$$n = 8 \Rightarrow v = 7 \Rightarrow t_c = 1.895$$

$$1.25 \pm \frac{4.200}{\sqrt{8}} \times 1.895$$

$$\Rightarrow \mu_d \in [-1.564, 4.064]$$

Find the degrees of freedom and use a GDC to find the characteristic value of t.
Use the formula $\bar{d} \pm \frac{s_d}{\sqrt{n}} \times t_c$

Method II

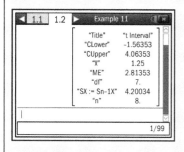

Store data into lists and find the differences. Apply the t-confidence interval with data in the difference list.

	1.1	1.2	►	Example 11	
	"Title"	"t Interval"			
	"CLower"	-1.56353			
	"CUpper"	4.06353			
	"x̄"	1.25			
	"ME"	2.81353			
	"df"	7.			
	"SX := Sn-1X"	4.20034			
	"n"	8.			

1/99

	1.1	1.2	►	Example 11	
	A a	B b	C	D	
				=a[]-b[]	
1	60	63	-3		
2	58	55	3		
3	47	42	5		
4	80	76	4		
5	35	29	6		

C =a[:]-b[:]

$$\Rightarrow \mu_d \in [-1.564, 4.064]$$

In the previous example, the 90% confidence interval contains both negative and positive values, so we cannot say at this level of confidence that there is a difference between the effects of drugs A and B. Sometimes in such a case we can do another hypothesis test that can give us more reasonable and conclusive results. You will learn more about this in Example 16.

Exercise 3D

1 Given two sets of data in the following tables, calculate the differences and find the $(1 - \alpha)$% confidence interval for the mean of the differences:

a

Set A	15	17	23	17	18	20	19	16	16	21
Set B	18	15	23	19	15	18	20	15	18	22

$\alpha = 0.01$;

b

Set C	98	103	102	88	96	105	110	93	102	106	99	85
Set D	121	115	104	96	110	112	123	102	108	109	111	103

$\alpha = 0.1$;

c

Set E	0.55	0.71	0.66	0.58	0.82	0.77	0.9	1.02
Set F	0.62	0.68	0.74	0.69	0.78	0.65	0.8	0.95

$\alpha = 0.05$.

2 Bob and Rick are two laboratory technicians and they measure haemoglobin levels (Hgb) in the same blood samples of 12 male patients. The normal level of that oxygen-transport metalloprotein is between 138 and 175 g/l. They obtained the following results in g/l:

Blood sample	1	2	3	4	5	6	7	8	9	10	11	12
Bob	144	153	170	183	125	95	148	177	160	155	170	135
Rick	141	161	173	174	119	104	135	175	164	158	167	142

a Find the differences between the results obtained by Bob and Rick.
b Calculate a 95% confidence interval of the mean of differences in results obtained by Bob and Rick.

3.3 Hypothesis testing

Setting up and testing hypotheses is an essential part of statistical studies. In order to formulate such a test, usually a theory has been proposed but has not yet been proved to be true. For example, suppose a claim has been made that a new drug to help fight infection works better than the current drug. We wish to set up and test a hypothesis to determine whether this claim is true.

Hypothesis testing always starts with a stated **claim** about a **population parameter**. The Mathematics Higher Level syllabus includes testing only for the population mean of a random normal distribution, and as such our study of this topic will be oriented thus.

In general, when we are discussing hypotheses we consider two statements that are directly contradictory to each other. The process of testing the hypotheses provides us with arguments as to why a certain hypothesis can be accepted or rejected.

The stated hypothesis is called the **null hypothesis** and we denote it by H_0 and the **alternative hypothesis** states the opposite and is denoted by H_1.

Let's consider again the example of broilers from section 3.2. Suppose that data from previous years suggests that the mean weight of the broiler population is 2 kg. We wish to estimate the mean weight of broilers this year, and to do so we take a sample of a certain size. The sample mean is found to be 2.10 kg.

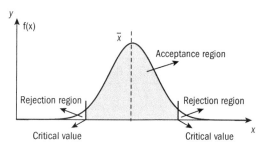

Two-tail test

The null hypothesis always states no change, i.e. that the mean weight of broilers this year is also 2 kg. We write this as $\left(H_0 : \mu = \overline{x}\right)$.

The alternative hypothesis can have different statements depending on the type of test we wish to perform.

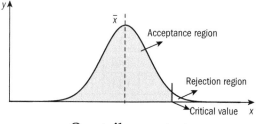

One-tail upper test

There are two types of hypothesis testing and three types of the alternative hypotheses:

i   **Two-tail test** ($H_1 : \mu \neq \overline{x}$) The mean weight is not equal to 2 kg.

ii **One-tail upper test** ($H_1 : \mu > \overline{x}$) The mean weight is more than 2 kg.

iii **One-tail lower test** ($H_1 : \mu < \overline{x}$) The mean weight is less than 2 kg.

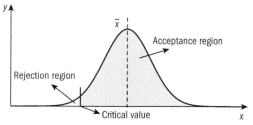

One-tail lower test

We must also decide which significance level, α, we require in order to conclude that a certain hypothesis is valid, with $(1 - \alpha)\%$ certainty.

The most common significant levels used are 1%, 5% and 10%.

Significace level is directly related to the confidence interval, so if we are 95% sure that the mean value is in the calculated confidence interval we can accept the null hypothesis at the 5% significant level of the test.

When calculating testing statistics, just as when calculating the confidence intervals in the previous section, we will use either z-statistics or t-statistics. We will use **z-statistics** when **the variance is known**, and **t-statistics** when **the variance is unknown** (regardless of the sample size).

There are two ways to make a decision based upon the calculated values.

The Critical value is the z-value or t-value found on the significance level of the test. If the calculated z- or t-value of the test lies outside the so-called acceptance region, we reject the null hypothesis, otherwise we do not reject it.

The **p-value** is the probability that the parameter we're investigating (i.e. the mean) lies within the rejection region, given that the null hypothesis is true. If the **p-value is greater** than the significance level we cannot say that we accept the null-hypothesis, but rather say that **'we have no sufficient evidence to reject'**, or simply **'fail to reject the null hypothesis'**.

We have **four steps** in hypothesis testing:

Step 1	State the null and alternative hypothesis;
Step 2	Set the criteria for a decision;
Step 3	Calculate the statistics;
Step 4	Decide upon calculated statistics and decision criteria.

Hypothesis testing for μ when σ is known

As in the calculation of confidence intervals, we are going to use z-statistics in hypothesis testing. Let's look at an example for which we have already calculated the confidence interval (see Example 4).

Example 12

In a certain country it is believed that the mean height of the male population is 182 cm and the standard deviation is 5 cm. A random sample of 100 men was taken from the population and the mean height was found to be 183.6 cm.

a State the null and alternative hypotheses;

b Use a two-tail test at the 10% significance level to decide whether or not the claim is true.

a H_0: "The mean height is 182 cm" ($\mu = 182$) H_1: "The mean height is not 182 cm" $\quad$ ($\mu \neq 182$)	*Step 1: State the null hypothesis that confirms the claim. State alternative hypothesis.*
b Method I $\mu = 182$, $\sigma = \dfrac{5}{\sqrt{100}} = 0.5$, $\bar{x} = 183.6$	*Step 2: Calculate the z-value.*
$z = \dfrac{\bar{x} - \mu}{\sigma} \Rightarrow z = \dfrac{183.6 - 182}{0.5} = 3.2$	
Either $\alpha = 0.1 \Rightarrow z_{\frac{\alpha}{2}} = \Phi^{-1}(0.05) = 1.645$	*Step 3: Find the z-critical value.*
$3.2 > 1.645 \Rightarrow z > z_{\frac{\alpha}{2}}$ We reject the null hypothesis.	*Step 4: Compare the z-critical value with the calculated value and make a decision.*
Or $P((180.4 > \bar{X}) \text{ or } (\bar{X} > 183.6)|\mu = 182)$	*Step 3: For the calculated z value find the corresponding p-value.*
$= 1 - P(180.4 < \bar{X} < 183.6|\mu = 182)$ $= 1 - P(-3.2 < Z < 3.2) = 0.00137$	
$0.00137 < 0.1$ we reject the null hypothesis at the 10% significant level.	*Step 4: Compare the p-value with the significant level and make a decision.*
Method II	*Use z-test statistics feature on a GDC*
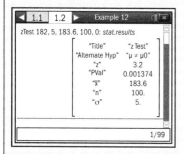	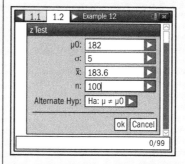
$0.00137 < 0.1$ we reject the null hypothesis.	*Compare the p-value with the significance level and make a decision.*

Due to the symmetrical properties of the confidence interval and the acceptance region in a two-tail test, this example highlights some important results:

- We found the 90% confidence interval for the mean of the population to be [182.8, 184.4] and we see that 182 is not in this interval. We therefore expected to reject the null hypothesis.
- Your calculator gives you the calculated z-value "z" = 3.2, but it is much simpler to make a decision based upon the p-value when it is available.
- When we compare the p-value with each of the common significance levels (1%, 5%, and 10%), we conclude that we reject the null hypothesis upon all three significance levels, since 0.00137 is smaller than each and every one of them.

Let's use another example for which we have already calculated the confidence interval.

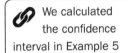 We calculated the confidence interval in Example 5

Example 13

After a rainy night, 12 worms have surfaced on sand. Their lengths, measured in cm, were as follows: 12.0, 11.1, 10.5, 10.8, 12.1, 10.4, 10.9, 12.2, 10.9, 11.9, 11.2, 11.6. It is known that the worms came from a population that follows a normal distribution with the mean value of 10.5 cm and the standard deviation of 2 cm. There is a belief that the worms are growing larger.

a State the null and alternative hypotheses;

b Use a one-tail upper test at the 5% significance level to decide whether or not the claim is true.

a H_0: "The mean length is 10.5 cm" $(\mu = 10.5)$ H_1: "The mean length is larger than 10.5 cm"$(\mu > 10.5)$	*Step 1: State the null hypothesis that confirms the claim. State the alternative hypothesis.*
b Method I $\mu = 10.5, \sigma = \dfrac{2}{\sqrt{12}} = 0.577, \bar{x} = 11.3$ $z = \dfrac{\bar{x} - \mu}{\sigma} \Rightarrow z = \dfrac{11.3 - 10.5}{0.577} = 1.38564$	*Step 2: Calculate the z-value.*
Either $\alpha = 0.05 \Rightarrow z_c = \Phi^{-1}(0.05) = 1.645$ $1.38564 < 1.645 \Rightarrow z < z_c$ We have no sufficient evidence to reject the null hypothesis.	*Step 3: Find the z-critical value.* *Step 4: Compare the z-critical value with the calculated value and make a decision.*
Or $P(\bar{x} > 11.3 \mid \mu = 10.5) = P(z > 1.38564)$ $= 0.0829$	*Step 3: For the calculated z-value find the corresponding p-value.*

Since 0.0829 > 0.05 we have no sufficient evidence to reject the null hypothesis at the 5% significance level.

Method II

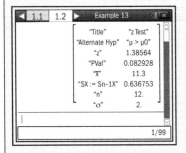

0.0829 > 0.05 we have no sufficient evidence to reject the null hypothesis at the 5% significance level.

Step 4: Compare the p-value with the significance level and make a decision.

Use z-test data feature on a GDC to find the p-value.

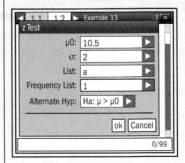

Compare the p-value with the significance level and make a decision.

Confidence interval analysis cannot be compared to the results obtained through one-tail hypothesis testing (upper or lower), since a confidence interval is symmetric about the mean, whereas an acceptance or rejection region for one-tail testing is not symmetric.

When we compare the *p*-value in Example 13 with the three common significance levels (1%, 5%, and 10%) it's clear that we reject the null hypothesis only upon the 10% significance level (0.0829 < 0.1), whilst upon the remaining two significance levels we have no sufficient evidence to reject the null hypothesis (since 0.0829 > 0.05 and 0.0829 > 0.01).

Exercise 3E

1 It is believed that a random sample of *n* observations is taken from the population with the mean value μ_0 and the standard deviation σ.
The sample has the mean value of $\bar{x}$. Given the hypotheses
$H_0 : \mu = \mu_0$, $H_1 : \mu \neq \mu_0$ test the claim at the α significance level if:

a $n = 20$, $\mu_0 = 10$, $\bar{x} = 12$, $\sigma = 2$ and $\alpha = 0.1$

b $n = 25$, $\mu_0 = 2$, $\bar{x} = 1.9$, $\sigma = 0.3$ and $\alpha = 0.05$

c $n = 119$, $\mu_0 = -235$, $\bar{x} = -238$, $\sigma = 12.8$ and $\alpha = 0.01$

2 It is believed that a random sample of *n* observations is taken from the population with the mean value μ_0 and the standard deviation σ.
The sample has the mean value of $\bar{x}$. Given the hypotheses
$H_0 : \mu = \mu_0$, $H_1 : \mu < \mu_0$ test the claim at the α significance level if:

a $n = 20$, $\mu_0 = 10$, $\bar{x} = 9$, $\sigma = 4$ and $\alpha = 0.1$

b $n = 50$, $\mu_0 = 21.4$, $\bar{x} = 21.2$, $\sigma = 0.75$ and $\alpha = 0.01$

c $n = 119$, $\mu_0 = -235$, $\bar{x} = -238$, $\sigma = 12.8$ and $\alpha = 0.01$

3 It is believed that a random sample of n observations is taken from the population with the mean value μ_0 and the standard deviation σ. The sample has the mean value of $\bar{x}$. Given the hypotheses $H_0 : \mu = \mu_0$, $H_1 : \mu > \mu_0$ test the claim at the α significance level if:

a $n = 20$, $\mu_0 = 10$, $\bar{x} = 12$, $\sigma = 5$ and $\alpha = 0.05$

b $n = 40$, $\mu_0 = 27.3$, $\bar{x} = 28.0$, $\sigma = 3.6$ and $\alpha = 0.1$

c $n = 92$, $\mu_0 = -73$, $\bar{x} = -71.6$, $\sigma = 3.72$ and $\alpha = 0.01$

4 After a rainy period, 15 snails have been harvested in the forest. Their weights, measured in g, were as follows: 22, 25, 31, 35, 28, 38, 25, 30, 35, 27, 29, 30, 34, 33, 28. It is known that snails in the forest came from a population that follows a normal distribution with the mean value of 26 g and the standard deviation of 3.5 g. There is a belief that the harvested snails are not from that population.

a State the null and alternative hypotheses.

b Use an appropriate test at the 1% significance level to decide whether or not the claim is true.

5 A company claims that the level of fat in 100 ml of a lactose free drink is 1.4 g with the standard deviation of 0.3 g. In a sample of eight drinks the following levels of fat were measured: 1.43, 1.52, 1.35, 1.38, 1.42, 1.46, 1.45, 1.55. There is a belief that the company is producing a lactose free drink with more fat.

a State the null and alternative hypotheses.

b Test at the 5% significance level whether or not the claim is true.

6 An eco-farm produces 100% pure apple juice. They package juice in bottles that they claim have a volume of 300 ml and a standard deviation of 7.3 ml. A box of 20 bottles is taken for inspection and the following volumes (in ml) were measured:

295, 288, 293, 301, 302, 285, 288, 290, 305, 300, 298, 299, 289, 304, 302, 290, 288, 300, 293, 305.

There is a belief that the bottles contain less volume than stated.

a State the null and alternative hypotheses.

b Test at the 10% significance level whether or not the claim is true.

Hypothesis testing for μ when σ is unknown

As with calculations of confidence intervals, if we do not know the standard deviation when hypothesis testing we use t-statistics, irrespective of the sample size.

> We have already found the confidence interval for this instance in Example 7.

Example 14

A chocolate company claims that the energy level in certain 100 g chocolates is 2500 kJ. In a sample of five 100 g chocolates the mean value of the energy level was found to be $\bar{x} = 2540$ kJ, whilst the estimate for population standard deviation was $s = 120$ kJ.
a State the null and alternative hypotheses;
b Test at the 1% significance level whether or not the claim is true.

a H_0: "The energy level is 2500 kJ" ($\mu = 2500$) H_1: "The energy level is not 2500 kJ" ($\mu \neq 2500$)	*Step 1: State the null hypothesis that confirms the claim. State the alternative hypothesis for a two-tail test.*
b **Method I** $t = \dfrac{\bar{x} - \mu}{\dfrac{s}{\sqrt{n}}} \Rightarrow t = \dfrac{2540 - 2500}{\dfrac{120}{\sqrt{5}}} = 0.7454$	*Step 2: Calculate the t-value.*
Either $n = 5 \Rightarrow v = 4, \alpha = 0.01 \Rightarrow t_c = 4.604$	*Step 3: Find the t-critical value.*
$0.7454 < 4.604 \Rightarrow t < t_c$ We have no sufficient evidence to reject the null hypothesis at the 1% significance level.	*Step 4: Compare the t-critical value with the calculated value and make a decision.*
Or $P((2460 > \bar{X})$ or $(\bar{X} > 2540)\mid\mu = 2500)$ $= 1 - P(2460 < \bar{x} < 2540\mid\mu = 2500)$ $= 1 - P(-0.74536 < t < 0.74536) = 0.497$	*Step 3: For the calculated t value find the corresponding p-value.*
$0.497 > 0.01$ we have no sufficient evidence to reject the null hypothesis at the 1% significant level.	*Step 4: Compare the p-value with the significant level and make a decision.*
Method II	*Use t-test stat feature on a GDC.*
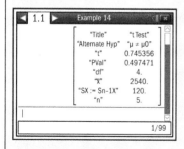	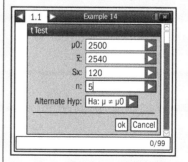
$0.497 > 0.01$ we have no sufficient evidence to reject the null hypothesis at the 1% significant level.	*Compare the p-value with the significance level and make a decision.*

Due to the symmetrical properties of the confidence interval and the acceptance region in a two-tail test, this example highlights some important results:

- We found the 90% confidence interval for the energy level in 100 g chocolates to be [2293, 2787] and we see that 2500 lies within this interval. Therefore, we do not have enough evidence to reject the null hypothesis.
- Your calculator gives you the calculated t-value "t" = 0.745, but it is much simpler to make a decision based upon the p-value.
- When we compare the p-value with each of the common significance levels (1%, 5%, and 10%), we conclude that we have no evidence to reject the null hypothesis upon every one of these three significance levels since 0.497 is larger than each.

We calculated the confidence interval for the following problem when we studied Example 9.

Example 15

In a box of six eggs, the weights (measured in grams) of each egg were as follows: 62, 63, 65, 62, 66, 66. The farmer claims that the mean weight of one of their eggs is 66 g. Test at the 5% significance level whether the eggs in the boxes have a mean weight less than 66 g.

H_0: "The mean weight is 66 g" ($\mu = 66$) H_1: "The mean weight is less than 66 g" ($\mu < 66$)	*Step 1: State the null hypothesis that confirms the claim. State alternative hypothesis for a one-tail lower test.*
Method I $\bar{x} = 64,\ s = 1.897$ $t = \dfrac{\bar{x} - \mu}{\dfrac{s}{\sqrt{n}}} \Rightarrow t = \dfrac{64 - 66}{\dfrac{1.897}{\sqrt{6}}} = -2.582$	*Step 2: Calculate the t-value.*
Either $n = 6 \Rightarrow v = 5, \alpha = 0.05 \Rightarrow t_c = -2.015$ $-2.582 < -2.015 \Rightarrow t < t_c$ We reject the null hypothesis at the 5% significant level.	*Step 3: Find the t-critical value.* *Step 4: Compare the t-critical value with the calculated value. Notice that t lies outside the acceptance region so we make a decision.*
Or $P(\bar{x} < 64 \mid \mu = 66)$ $P(-2.582 < t) = 0.02466$	*Step 3: For the calculated t-value find the corresponding p-value.*
$0.02466 < 0.05$ therefore we reject the null hypothesis at the 5% significant level.	*Step 4: Compare the p-value with the significance level and make a decision.*

Method II

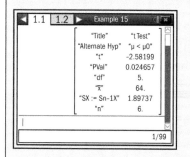

Use t-test stat feature on a GDC to find the p-value.

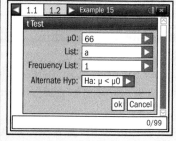

$0.024657 < 0.05$ therefore we reject the null hypothesis that the mean weight is $66\,\text{g}$ at the 5% significance level.

As with z-testing, confidence interval analysis for a t-distributed variable cannot be compared to the results obtained through one-tail hypothesis t-testing (upper or lower), since a confidence interval is symmetric about the mean, whereas an acceptance or rejection region for one-tail t-testing is not symmetric.

Exercise 3F

1 A random sample of n observations is taken from the population with the mean value μ_0. The sample has the mean value of $\bar{x}$ and an unbiased estimate of the population standard deviation s. Given the hypotheses $H_0 : \mu = \mu_0$, $H_1 : \mu \neq \mu_0$ test the claim at the α significance level if:

a $n = 10$, $\mu_0 = 5$, $\bar{x} = 4.8$, $s = 1.3$ and $\alpha = 0.01$

b $n = 25$, $\mu_0 = 2$, $\bar{x} = 1.9$, $s = 0.3$ and $\alpha = 0.1$

c $n = 7$, $\mu_0 = -36$, $\bar{x} = -35.3$, $s = 0.523$ and $\alpha = 0.05$

2 It is believed that a random sample of n observations is taken from the population with the mean value μ_0. The sample has a mean value of $\bar{x}$ and the unbiased estimate of the population standard deviation is s. Given the hypotheses $H_0 : \mu = \mu_0$, $H_1 : \mu < \mu_0$ test the claim at the α significance level if:

a $n = 30$, $\mu_0 = 15$, $\bar{x} = 14.2$, $s = 2.2$ and $\alpha = 0.05$

b $n = 10$, $\mu_0 = 122$, $\bar{x} = 119.8$, $s = 2.32$ and $\alpha = 0.01$

c $n = 6$, $\mu_0 = 627$, $\bar{x} = 622.8$, $s = 12.6$ and $\alpha = 0.1$

3 It is believed that a random sample of n observations is taken from the population with the mean value μ_0. The sample has the mean value of $\overline{x}$ and the unbiased estimate of the population standard deviation s. Given the hypotheses $H_0 : \mu = \mu_0$, $H_1 : \mu > \mu_0$ test the claim at the α significance level if:

a $n = 20$, $\mu_0 = 1$, $\overline{x} = 0.95$, $s = 0.335$ and $\alpha = 0.1$;

b $n = 8$, $\mu_0 = 25$, $\overline{x} = 26.4$, $s = 1.12$ and $\alpha = 0.05$;

c $n = 15$, $\mu_0 = 754$, $\overline{x} = 758.6$, $s = 14.2$ and $\alpha = 0.01$.

4 An ice-cream factory claims that the average volume of a particular ice-cream product is 120 ml. There are eight ice-creams in a box and their volumes in ml are as follows:
119, 123, 121, 120, 118, 116, 123, 122. State the hypotheses and test at the 1% significance level whether or not the factory advertised the correct volume.

5 A manufacturer claims that the life expectancy of an LED lamp is 30 000 hours. A random sample of six LED lamps is tested and the following data is obtained:

29 500   28 350   30 300   30 250   29 350   29 600

State the hypotheses and test at the 10% significance level whether or not the manufacturer claims a longer life expectancy of the LED lamps.

> **?** In some countries the comma is used every three decimal places for making numbers with many digits easier to read. In other countries, a space is used instead of the comma. In some computer programming languages, an underscore is used.

Significance testing for matched pairs

In Example 11, we studied how to use confidence intervals to compare the data obtained in a matched pairs study. Example 16 will illustrate how to use significance testing to compare the same data.

Example 16

A group of patients at a hospital suffer from chronic iron deficiency. They are treated with two new drugs: drug A and drug B. First, they are treated with drug A and, after a few days, when the influence of drug A has worn off, they are treated with drug B. A serum iron test is conducted and the results (in $\mu g / dl$) are given in the following table:

Patient	a	b	c	d	e	f	g	h
Drug A	60	58	47	80	35	55	53	40
Drug B	63	55	42	76	29	61	51	41

a Find the differences between the results obtained by drug A and drug B;

b State the hypotheses and test at the 10% significance level whether or not there is a difference between the effects of these two drugs.

<table>
<tr><td>**a**</td><td>$d_i = A - B$</td><td>-3</td><td>3</td><td>5</td><td>4</td><td>6</td><td>-6</td><td>2</td><td>-1</td></tr>
</table>

Subtract the results of drug B from the results of drug A.

b H_0: "There is no difference in drug effect"

$(\mu_d = 0)$

H_1: "There is a difference in drug effect"

$(\mu_d \neq 0)$

Step 1: State the null hypothesis that confirms the claim. State alternative hypothesis for a two-tail test.

Method I

$\bar{d} = 1.25, \; s_d = 4.200$

$$t = \frac{\bar{d} - \mu}{\dfrac{s_d}{\sqrt{n}}} \Rightarrow t = \frac{1.25 - 0}{\dfrac{4.200}{\sqrt{58}}} = 0.842$$

Step 2: Calculate the t-value.

Either

$n = 8 \Rightarrow v = 7, \; \alpha = 0.1 \Rightarrow t_c = 1.895$

$|0.842| < |1.895| \Rightarrow |t| < |t_c|$

Step 3: Find the t-critical value.

Step 4: Compare the t-critical value with the calculated value and make a decision.

We have no sufficient evidence to reject the null hypothesis.

Or

$P((-1.25 > \bar{d}) \text{ or } (\bar{d} > 1.25)|\mu = 0)$

$= 1 - P(-1.25 < \bar{d} < 1.25|\mu = 0)$

$= 1 - P(-0.842 < t < 0.842) = 0.428$

Step 3: For the calculated t-value find the corresponding p-value.

0.428 > 0.1 we have no sufficient evidence to reject the null hypothesis at the 10% significance level.

Step 4: Compare the p-value with the significance level and make a decision.

Method II

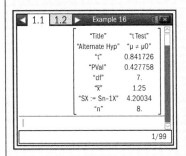

Use a GDC to store data into lists and find the differences. Apply the t-test with data in the difference list.

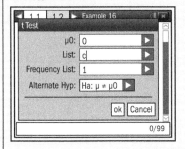

0.428 > 0.1 we have no sufficient evidence to reject the null hypothesis at the 10% significance level.

Compare the p-value with the significance level and make a decision.

In Example 11, we saw that the 90% confidence interval $[-1.564, 4.064]$ contains both negative and positive values. We had concluded that we could not say at this level of confidence there was a difference between the effects of the two drugs. However, significance testing gives us more information and we determine the probability that the mean lies in the acceptance region.

Exercise 3G

1 The times (in seconds) that it took eight players to solve two Rubik's Cube puzzles are shown in the table below.

Player	A	B	C	D	E	F	G	H
Cube 1	22	35	41	30	28	46	52	36
Cube 2	26	38	40	34	30	44	48	28

Test at a 5% significance level whether or not there is a difference between the finishing times on the two different cubes.

2 Six darts players are testing darts that have a radical new flight design. Their scores (out of 100) are given in the table below.

Player	A	B	C	D	E	F
Dart–old design	85	92	100	97	89	91
Dart–new design	90	95	98	99	93	92

Test at a 10% significance level whether or not there is a difference between the old and the new design of dart.

3 A personal trainer at a fitness club claims that members will lose weight after taking his aqua aerobics classes. A group of 12 students decided to join the programme. The table shows their weights, given in kg, before and after a 30-day period.

Student	A	B	C	D	E	F	G	H	I	J	K	L
Weight before the class	55	82	63	69	65	58	88	64	72	75	90	77
Weight after the class	52	84	61	69	62	57	84	66	70	71	94	77

Test at a 5% significance level whether or not the personal trainer's advertising is fair.

3.4 Type I and Type II errors

In this section we are going to discuss what kind of errors we can make in the statistical testing we have studied. Let's start by noting that every **null hypothesis** can in fact have two possible values if we think of them as statements: **true** or **false**.

- We say that we make a **Type I error** when we **reject a true null hypothesis**.
- We make a **Type II error** when we **fail to reject a false null hypothesis**.

A real-life example of these two types of errors can be seen in a court trial. Sometimes an innocent person is convicted, whilst other times a guilty person walks free. In a democracy, as everyone is presumed innocent at the beginning of a trial (our null hypothesis), we say that:

- Convicting an innocent person corresponds to a Type I error
- Freeing a guilty person corresponds to a Type II error.

In the table below α is the probability of rejecting a true null hypothesis and β is the probability of failing to reject a false null hypothesis.

> The probability of making a Type I error or the area of the rejection region is denoted by α. The probability of making a Type II error is denoted by β. The value of $1 - \beta$ is called the power of the test.

		Reality	
		Null hypothesis is true	Null hypothesis is false
Decision based on collected data	Fail to reject null hypothesis	Correct decision $1-\alpha$	Type II error (Failing to reject false hypothesis) β
	Reject null hypothesis	Type I error (Rejecting true hypothesis) α	Correct decision $1-\beta$

> **?** In medicine, researchers perform a lot of statistical testing. In this area, a **Type I error** is known as **false positive** whilst a **Type II error** is known as **false negative**. For example we perform a test to ascertain whether a patient is infected with a certain virus. The false positive means that the test has shown that the patient does have a viral infection but they are actually not infected by the virus. A false negative means that the test hasn't shown viral presence in the patient's body, but they actually are infected by the virus. If we are wondering which type of error is better to make, we can see from this medical example that although a false positive might cause a patient to worry, eventually due to the lack of viral symptoms, combined with results of different tests, the patient will find out that nothing is wrong with their health. On the other hand, a false negative will give a patient a false assurance that nothing is wrong with their health, and since the virus will not be treated, it can have serious consequences for the health of the patient. We can thus conclude that a Type II error is *less* desirable than a Type I error.

Suppose we roll a die 20 times and do not roll a six.

Our null hypothesis is that the die is fair, and the alternative hypothesis is that the die is biased.

Even though the probability of obtaining no sixes in 20 rolls of the die is fairly small ($p = 0.0261$), it is certainly possible.

- A Type I error might occur when the die is actually fair (i.e. the null hypothesis is true), but we reject the null hypothesis and conclude that the die is biased.
- A Type II error might occur when the die is indeed biased (i.e. the null hypothesis is false), but we accept the null hypothesis and conclude that the die is fair.

Example 17

There are two coins. One coin is fair and one coin is biased so that the probability of obtaining a "head" with this coin is $p = \dfrac{2}{3}$. We take one of these two coins and flip it four times.

The variable X denotes the number of "heads" on the fair coin and the variable Y denotes the number of "heads" on the biased coin.

a Find probability distribution tables for both coins.

The null hypothesis H_0 is "the selected coin is fair" whilst the alternative hypothesis H_1 is "the selected coin is biased". We decide that we are going to reject the null hypothesis if we obtain four "heads".

b Find the probability of obtaining a Type I error;

c Find the probability of obtaining a Type II error.

a Fair coin:

$X = x$	0	1	2	3	4
$P(X = x)$	$\dfrac{1}{16}$	$\dfrac{4}{16}$	$\dfrac{6}{16}$	$\dfrac{4}{16}$	$\dfrac{1}{16}$

Use the binomial PDF:

$$P(X = x) = \binom{4}{x}\left(\frac{1}{2}\right)^{4-x}\left(\frac{1}{2}\right)^{x} = \frac{1}{16}\binom{4}{x}$$

Biased coin:

$Y = y$	0	1	2	3	4
$P(Y = y)$	$\dfrac{1}{81}$	$\dfrac{8}{81}$	$\dfrac{24}{81}$	$\dfrac{32}{81}$	$\dfrac{16}{81}$

Use the binomial PDF:

$$P(Y = y) = \binom{4}{y}\left(\frac{1}{3}\right)^{4-y}\left(\frac{2}{3}\right)^{y} = \binom{4}{y}\frac{2^y}{3^4}$$

b $\alpha = P\left(X = 4 \mid H_0\right) = \dfrac{1}{16}$

Type I error is when we reject the null hypothesis when it is actually true.

c $\beta = P\left(Y \neq 4 \mid H_1\right) = 1 - \dfrac{16}{81} = \dfrac{65}{81}$

Type II error is when we fail to reject the null hypothesis when it is false.

Notice that we could have decided to reject the null hypothesis for a different event, for instance if we obtain no "heads". In that case the probability of making a Type I error will remain the same $P(X = 0 \mid H_0) = \dfrac{1}{16}$, but the probability of making a Type II error will increase, $P(Y \neq 0 \mid H_1) = 1 - \dfrac{1}{81} = \dfrac{80}{81} > \dfrac{65}{81}$, as expected.

Since the probability of making a Type II error is smaller in example 17 than it is in the instance of obtaining no "heads", we can conclude that the test where we reject the null hypothesis when we obtain four heads is a better test than when we reject the null hypothesis after obtaining no heads.

Example 18

Let's assume that $X \sim Po(30)$ and we test this null hypothesis against the hypothesis that the parameter of the distribution is greater than 30. The acceptance region for the null hypothesis contains all the x values less than or equal to 35.

a   Find the probability of a Type I error;

b   It was found that the actual value of the parameter was 40. Find the probability of a Type II error;

c   If we expand the acceptance region to all the x values less than or equal to 38, find the probabilities for both Type I and II errors. What can you conclude?

a $\alpha = P(X \geq 36 | m = 30)$

$= 1 - P(X \leq 35 | m = 30) = 0.157$

$H_0 : X \sim Po(30)$

Use a GDC to find the probability.

b $\beta = P(X \leq 35 | m = 40) = 0.242$

$H_1 : X \sim Po(40)$

Use a GDC to find all the probabilities.

c $\alpha' = P(X \geq 39 | m = 30)$

$= 1 - P(X \leq 38 | m = 30) = 0.0648$

$\beta' = P(X \leq 38 | m = 40) = 0.416$

We notice that Type I and Type II errors are connected, so when we decrease a Type I error we increase a Type II error.

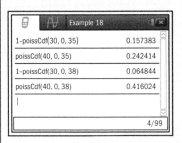

1-poissCdf(30, 0, 35)	0.157383
poissCdf(40, 0, 35)	0.242414
1-poissCdf(30, 0, 38)	0.064844
poissCdf(40, 0, 38)	0.416024

4/99

$\alpha > \alpha' \Rightarrow \beta < \beta'$

Example 19

Let's assume that $X \sim B\left(100, \frac{1}{4}\right)$ and we test this null hypothesis against the hypothesis that the probability $p \neq \frac{1}{4}$. The acceptance region for the null hypothesis contains all the x values that are within 8 of the expected value of X.

a   Find $E(X)$ and the probability of a Type I error;

b   It was found that the actual value of the probability was $p = \frac{2}{5}$. Find the probability of a Type II error;

c   If we expand the acceptance region of the null hypothesis to all the x values that are within 6 of the expected value, find the probabilities for both Type I and II errors. What can you conclude?

a $E(X) = 100 \times \dfrac{1}{4} = 25$

$\alpha = P\left((X \leq 16) \text{ or } (X \geq 34) \middle| p = \dfrac{1}{4} \right)$

$= 1 - P\left(17 \leq X \leq 33 \middle| p = \dfrac{1}{4} \right) = 0.0487$

b $\beta = P\left(17 \leq X \leq 33 \middle| p = \dfrac{2}{5} \right) = 0.0913$

c $\alpha' = P\left((X \leq 18) \text{ or } (X \geq 32) \middle| p = \dfrac{1}{4} \right)$

$= 1 - P\left(19 \leq X \leq 31 \middle| p = \dfrac{1}{4} \right) = 0.132$

$\beta' = P\left(19 \leq X \leq 31 \middle| p = \dfrac{2}{5} \right) = 0.0398$

Again we notice that Type I and Type II errors are connected, but this time when we increase a Type I error we decrease a Type II error.

Apply the formula for binomial distribution E(X) = np.

$H_0: X \sim B\left(100, \dfrac{1}{4} \right)$

Use a GDC to find the probability.

$H_1: X \sim B\left(100, \dfrac{2}{5} \right)$

Use a GDC to find all the probabilities.

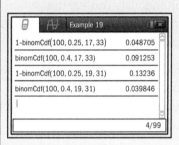

Example 19	
1-binomCdf(100, 0.25, 17, 33)	0.048705
binomCdf(100, 0.4, 17, 33)	0.091253
1-binomCdf(100, 0.25, 19, 31)	0.13236
binomCdf(100, 0.4, 19, 31)	0.039846
	4/99

$\alpha < \alpha' \Rightarrow \beta > \beta'$

The calculations we performed in Examples 18 & 19 can be seen visually in the following two normal distribution graphs.

Despite the distributions not being normal in Examples 18 & 19, we can still approximate both binomial and Poisson distributions by a normal distribution.

Normal distribution graph for a one-tail test

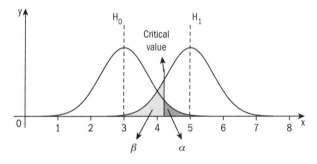

The diagram demonstrates that by moving the critical value vertical line

- to the right, we decrease Type I error but at the same time we increase Type II error;
- to the left, we increase Type I error but at the same time we decrease Type II error.

Two-tail test

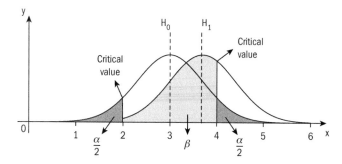

Again, this diagram demonstrates that by moving the critical value vertical lines simultaneously (notice that when we move one vertical critical value line to the left, the other one is moved to the right, so that the critical values are symmetric about H_0) the following occurs:

- When we decrease Type I error, we simultaneously increase Type II error;
- When we increase Type I error, we simultaneously decrease Type II error.

Exercise 3H

1 We believe that a normal random variable, X, is distributed such that $X \sim N(5, 0.4^2)$. We test this null hypothesis against the hypothesis that the mean $\mu \neq 5$. The acceptance region for the null hypothesis is $\{4.2 \leq X \leq 5.8\}$

   a Find the probability of Type I error.
   b It was found that the actual mean value was $\mu = 4.5$.
      Find the probability of Type II error.

2 Let's assume that $X \sim B\left(50, \dfrac{1}{2}\right)$ and we test this null

   hypothesis against the hypothesis that the probability $p > \dfrac{1}{2}$.

   The acceptance region for the null hypothesis is $\{X \leq 35\}$.
   a Find $E(X)$ and the probability of Type I error.
   b It was found that the actual value of the probability
      was $p = \dfrac{4}{7}$. Find the probability of Type II error.

3 Let's assume that $X \sim Po(45)$ and we test this null hypothesis against the hypothesis that the parameter of the distribution is less than 45. The acceptance region for the null hypothesis is $\{X \leq 52\}$.

   a Find the probability of Type I error.
   b It was found that the actual value of the parameter was 40.
      Find the probability of Type II error.

Review exercise

1 A box of 20 salmon is bought from a fish farm. The mean weight of a salmon was found to be 652 g and the farm owner states that the standard deviation of salmon he produces is 114 g.

   a Find the 95% confidence interval for the mean weight of salmon produced at the farm.

   b Find the 99% confidence interval for the mean weight of salmon produced at the farm.

   c Comment on the significance level and the width of the confidence interval.

2 In a medical laboratory the level of potassium in blood is measured by two types of biochemical analyzers. The range of the potassium level in the blood of a healthy person is between 270 and 390 mg/dl. The table below lists the measured levels of potassium, given in mg/dl, for 10 patients.

Patient	A	B	C	D	E	F	G	H	I	J
Analyzer I	235	352	410	280	341	325	428	388	272	310
Analyzer II	237	343	416	272	336	329	413	396	265	315

   a Find the differences between the results obtained by the two analyzers.

   b State the hypothesis and test at a 1% significance level whether or not there is a difference in measurement of the two types of biochemical analyzers.

3 Fifteen independent observations of a random sample are taken from a normal population. The sample gave the results $\sum_{i=1}^{15} x_i = 80$ and $\sum_{i=1}^{15} x_i^2 = 488$.

   a Calculate the unbiased estimates of the mean and variance for the given observations.

   b Find a 99% confidence interval for the population mean.

   c Interpret the meaning of the confidence interval at the given level.

4 The measurement of six independent random measurements may be assumed to follow a normal distribution with the mean value μ and the variance $\sigma^2 = 25$. Given that the confidence interval for the mean is found to be [47.2, 55.2], find:

   a the mean value of the sample;

   b the confidence level of the interval.

5 An automotive instrument manufacturing company claims that their speedometer shows the exact speed at which the car is driving. A random sample of ten cars was taken and they were tested on part of a straight 1 km racing track with the autopilot set at 120 km/h.

   a If the claim is true, how long would it take each car to travel the 1 kilometer track?

The times measured in seconds were as follows: 31.2, 30.8, 30.4, 30.8, 31.3, 32.1, 30.3, 31.4, 30.9, and 30.5.

   b Calculate the unbiased estimate of the mean and variance of the measured times.

An auto magazine claims that due to safety reasons the company deliberately sets up the speedometers to show a higher speed.

   c State the hypothesis and test the claim at the 5% significance level.

6 The measurement of n independent random measurements may be assumed to follow a normal distribution with the unbiased estimate of population variance $s^2 = 144$. Given that the 95% confidence interval for the mean is found to be [204, 216] find:

   a the mean value of the sample;
   b the sample size n.

7 A radar records the speed, v, in kilometres per hour, of bicycles on a bicycle lane. The speed of these bicycles is normally distributed. The results for 150 bicycles are recorded in the following table.

Speed	Number of bicycles
$0 \leq v < 10$	9
$10 \leq v < 20$	56
$20 \leq v < 30$	47
$30 \leq v < 40$	25
$40 \leq v < 50$	13

   a For the bicycles on the bicycle lane, calculate
      i an unbiased estimate of the mean speed;
      ii an unbiased estimate of the standard deviation of the speed.
   b For the bicycles on the bicycle lane, calculate
      i a 95% confidence interval for the mean speed;
      ii a 90% confidence interval for the mean speed.
   c Explain why one of the intervals found in part **b** is a subset of the other.

8 A population follows a normal distribution with the following parameters $N(\mu, \sigma^2 = 2)$. Let's assume that $\mu = 10$ to test the following hypotheses:

$H_0 : \mu = 10$

$H_1 : \mu < 10$,

using the mean of a sample of size 5.

a Find the appropriate critical regions corresponding to a significance level of

  i 0.1   **ii** 0.05

b If the actual population mean is 9.3, calculate the probability of making a Type II error when the level of significance is

  i 0.1   **ii** 0.05

c Explain the change in the probability of a Type I error related to the change in the probability of a Type II error.

Chapter 3 summary

Estimator and estimate

A random variable T is called an **unbiased estimator** for the population parameter θ if $E(T) = \theta$.

A specific **value** of that random variable is called an ***estimate***.

Given two estimators T_1 and T_2 of the population we say that T_1 is a **more efficient estimator** than T_2 if $\mathrm{Var}(T_1) < \mathrm{Var}(T_2)$.

$\overline{X} = \sum_{i=1}^{n} \dfrac{X_i}{n}$ is unbiased estimator for μ. $S^2 = \sum_{i=1}^{n} \dfrac{\left(X_i - \overline{X}\right)^2}{n-1}$ is unbiased estimator for σ^2.

$$S^2 = \dfrac{\sum_{i=1}^{n} x_i^2 - n(\overline{x})^2}{n-1} \text{ or } s^2 = \dfrac{n}{n-1}\sigma^2$$

Confidence interval for mean μ of population

i) When the population standard deviation σ is known $\left[\overline{x} - \dfrac{\sigma}{\sqrt{n}} \times z_{\frac{\alpha}{2}}, \overline{x} + \dfrac{\sigma}{\sqrt{n}} \times z_{\frac{\alpha}{2}}\right]$

ii) When the population standard deviation σ is unknown $\left[\overline{x} - \dfrac{s}{\sqrt{n}} \times t_c, \overline{x} + \dfrac{s}{\sqrt{n}} \times t_c\right]$

where $t_c = invt\left(1 - \dfrac{\alpha}{2}, v\right)$

Confidence interval for matched pairs

Calculate the differences between the observations and then find the confidence interval for the mean of differences.

Hypothesis testing

There are two types of hypothesis testing and three types of the alternative hypotheses:

i) **Two-tail test** ($H_1 : \mu \neq \overline{x}$)

ii) **One-tail upper test** ($H_1 : \mu > \overline{x}$)

iii) **One-tail lower test** ($H_1 : \mu < \overline{x}$)

There are **four steps** in hypothesis testing:

Step 1 State the null and alternative hypothesis;
Step 2 Set the criteria for a decision;
Step 3 Calculate the statistics;
Step 4 Decide upon calculated statistics and decision criteria.

Hypothesis testing for μ when σ is known

We use z-statistics in hypothesis testing.

Hypothesis testing for μ when σ is unknown

We use t-statistics in hypothesis testing regardless the sample size.

Significance testing for matched pairs

We calculate the differences between the observations and then use *t*-statistics in hypothesis testing regardless the sample size.

Type I and Type II errors

- We say that we make a **Type I error** when we **reject a true null hypothesis**.
- We make a **Type II error** when we **fail to reject a false null hypothesis**.

		Reality	
		Null hypothesis is true	Null hypothesis is false
Decision based on collected data	Fail to reject null hypothesis	Correct decision $1 - \alpha$	Type II error (Failing to reject false hypothesis) β
	Reject null hypothesis	Type I error (Rejecting true hypothesis) α	Correct decision $1 - \beta$

One-tail test

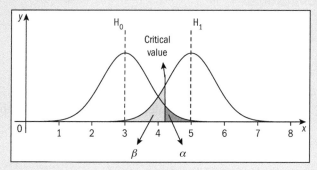

Two-tail test

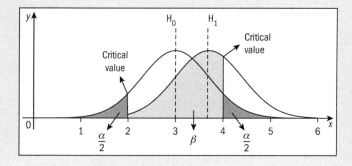

4 Statistical modeling

CHAPTER OBJECTIVES:

7.7 Introduction to bivariate distributions; covariance and (population) product moment correlation coefficient ρ; proof that $\rho = 0$ in the case of independence and $\rho = \pm 1$ in the case of a linear relationship between X and Y; definition of product moment correlation coefficient R in terms of n paired observations on X and Y. Its application to the estimation of ρ; informal interpretation of r, the observed value of R and scatter diagrams; use of the t-statistic to test the null hypothesis $\rho = 0$; knowledge of the facts that the regression of X on Y $(E(X)|Y = y)$ and Y on X $(E(Y)|X = x)$ are linear; least-squares estimates of these regression lines; the use of these regression lines to predict the value of one of the variables given the value of the other.

Before you start

You should know how to:

1 Use Expectation Algebra, e.g. If X and Y are two independent random variables, find:
$E(2X - Y) = 2E(X) - E(Y)$;
$Var(2X - Y) = 2^2Var(X) + Var(Y)$
$= 4Var(X) + Var(Y)$.

2 Find the mean and variance of a data set using the formulae, e.g. find the mean and variance of the following variables:

X	Y
5.7	−2.1
7.1	−3.3
2.3	0.9
3.9	1.2

$$\bar{x} = \frac{\sum_{i=1}^{n} x_i}{n};\ Var(X) = \frac{\sum_{i=1}^{n}(x_i - \bar{x})^2}{n} = \sum_{i=1}^{n}\frac{x_i^2}{n} - \bar{x}^2$$

$$\bar{x} = \frac{5.7 + 7.1 + 2.3 + 3.9}{4} = 4.75$$

$$\frac{\sum x^2}{n} = \frac{5.7^2 + 7.1^2 + 2.3^2 + 3.9^2}{4} = 25.85$$

$Var(X) = 25.85 - 4.75^2 = 1.81$

Similarly, $\bar{y} = -0.825$, $Var(Y) = 1.93$

3 Use the t-statistic to calculate probabilities, e.g. given that $X \sim t(v = 6)$, use the GDC to find the following probabilities:

a $P(X \le 1.2) = 0.862$ **b** $P(X \ge -0.52) = 0.689$
c $P(0.3 \le X \le 2.5) = 0.364$

Skills check:

1 If X, Y and Z are three standard independent normal random variables, find:
a $E(2Z - 3Y + 2X)$ **b** $Var(2Z - 3Y + 2X)$
c $E(XYZ)$

2 a Using the formulae, find the mean and variance of X and Y, as shown in the table.

X	Y
33	−21
49	−44
50	−23
42	−39

b Using a GDC, confirm your values for part **a**.

> On a GDC, you can enter the data in two different lists, and then select from the Stats calculation menu, '2-Variable Statistics', and see the values for both variables on the same screen.

3 Given that $X \sim t(v)$ use the GDC to find the following probabilities if:
a $v = 2$, $P(-0.4 \le X \le 0.8)$
b $v = 10$, $P(X \le 1.83)$
c $v = 7$, $P(-1.75 \le X)$
d $v = 20$, $P(-1.14 \le X \le 1.14)$

Bivariate distributions

Today it is a generally accepted fact that smoking can lead to lung cancer. In the USA, it is estimated that 90% of lung cancer deaths in males and about 75% in females are due to smoking. However, 150 years ago lung cancer was a relatively rare disease, comprising only about 1% of all malignant tumors seen in autopsies. An increase was seen throughout the 20th century, particularly after World War I, and mostly in males. A slower but equally steady increase in female incidence was also observed.

Initially it was thought that exposure to toxic gases released during the war, air pollution caused by increasing industrialization (including the automobile), and other such factors were causing the increase in lung cancer. However, countries with few effects from either the war or industrialization were also showing an increase in lung cancer incidence.

In attempting to discover the cause of this worldwide increase in the disease, a common factor emerged: there was also a worldwide increase in the use of tobacco. However, the general public, and in particular smokers and tobacco companies, were not ready to accept such a connection without strong scientific evidence.

In this chapter we will examine ways of handling data associated with two variables, in order to determine the nature of the relationship between them. We will also be studying measures of the degree to which a change in one variable effects a change in the other, i.e. two variables are positively related if one increases as the other increases, and two variables are negatively related if one decreases as the other increases. We will look at the measures of covariance and correlation which indicate *how* two variables are related (i.e. whether they are positively or negatively related) and also at the measure of correlation which describes the *degree* to which two variables are positively or negatively related.

We will begin by considering the relationship of two variables using their observed, or experimental, values.

4.1 Correlation

We are now going to consider two random variables, X and Y, measured on the same population. Each of these variables has a distribution of values. The n paired observations of X and Y form the joint distribution, X_i and Y_i, and the graph of the points (x_i, y_i) is called a scatter diagram. Each individual observation contains an x-value and a y-value, hence the data obtained is a set of n paired observations.

For example, let us consider the widely held assumption that students who perform well in Mathematics also perform well in Physics, and vice versa. We will compare the scores of ten students in both a Mathematics test and a Physics test in order to examine the relationship between these two variables. From the data in the table below we will draw a scatter graph by plotting each student's score in Mathematics against their score in Physics. The table below shows the corresponding values.

Student	A	B	C	D	E	F	G	H	I	J
X-Mathematics score	41	37	38	39	49	47	42	34	36	48
Y-Physics score	36	20	31	24	37	35	42	26	27	29

The following scatter diagram graphically displays the Mathematics and Physics scores of the ten students recorded in the table above.

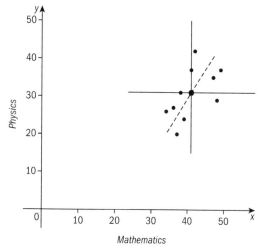

Horizontal and vertical lines that cross at the mean score for Mathematics (41) and Physics (31) have been added to the diagram, creating four quadrants.

As the majority of points are in the 3rd and 1st quadrants of this new translated coordinate system, this shows us that 'above average' scores in Mathematics usually correspond to 'above average' scores in Physics, and 'below average' scores in Mathematics usually correspond to 'below average' scores in Physics. The diagram therefore indicates that there is a positive relationship between a student's performance in Mathematics and Physics exams. If we allow a line to follow the general trend of the points, it would have a positive gradient. We say therefore that these two variables have a **positive correlation**.

Let us now consider the scores for ten students in both a Mathematics test and a History test, and draw a corresponding scatter diagram.

Student	K	L	M	N	O	P	Q	R	S	T
X-Mathematics score	4	22	4	45	22	34	5	27	17	26
Y-History score	54	17	10	20	26	12	35	15	40	25

The following scatter diagram graphically displays the scores of the ten students recorded in the table above.

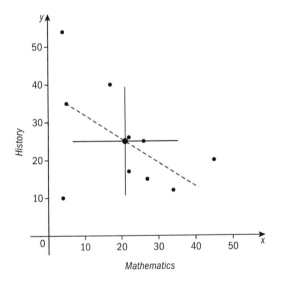

Again, a vertical and horizontal line are shown passing through the mean point (21, 25). Here the majority of points are in the 2nd and 4th quadrants. This shows us that 'above average' scores in Mathematics usually correspond to 'below average' scores in History, and 'above average' scores in History usually correspond to 'below average' scores in Mathematics.

The diagram therefore indicates that there is a negative relationship between a student's performance in Mathematics exams compared with history History exams. If we allow a line to follow the general trend of the points, it would have a negative gradient. We say therefore that these two variables have a **negative correlation**.

Below is the table of values for ten students' scores in a Mathematics exam and a Photography exam, followed by a scatter diagram showing this data.

Student	K	L	M	N	O	P	Q	R	S	T
X-Mathematics score	20	22	30	45	22	45	15	27	45	26
Y-Photography score	20	48	10	20	12	38	50	35	18	25

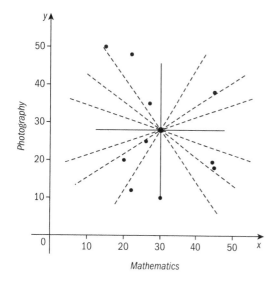

Here we cannot see a clear positive or negative relationship between a student's test score in Mathematics compared with their test score in Photography. It is difficult to see a line following the general trend of the points. We therefore say that these two variables have **no correlation**.

The first two graphs indicated a possible correlation between two variables, but they did not reliably tell us the strength of the correlation. We therefore need to calculate the **correlation coefficient**, which assesses the degree of the correlation between the two variables.

To do this, we will redraw the scatter diagrams that showed correlation, but this time the axes will be the lines drawn through the mean scores i.e. the origin will be the mean point.

This new scatter diagram shows the deviation $d_x = x - \bar{x}$ (the horizontal distance to the vertical axis) of the Mathematics score from the mean Mathematics score and the deviation $d_y = y - \bar{y}$ (the vertical distance to the horizontal axis) of the Physics score from the mean Physics score.

For student A, therefore, $d_x = 41 - 41 = 0$, and $d_y = 36 - 31 = +5$; hence, student A is represented on the graph by the point (0, 5). We do this for each of the students A through J.

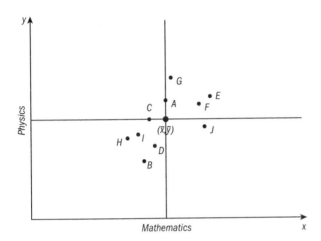

We have already seen in the first two scatter diagrams that a positive correlation exists if most of the points are either in the 1st and 3rd quadrants. Since we have redrawn the diagram with the mean as the origin, note that most of the products $d_x d_y$ will be positive, i.e. in the 1st quadrant both d_x and d_y are positive, and so their product will be positive, and in the 3rd quadrant, both d_x and d_y are negative, and so their product will also be positive.

Likewise, we have seen that a negative correlation exists if most of the points are either in the 2nd and 4th quadrants, and most of the products, $d_x d_y$, will therefore be negative.

The 3rd scatter diagram showed no correlation, since the points are more or less uniformly distributed in all 4 quadrants. In this case, about half of the products $d_x d_y$ will be positive, and about half will be negative.

The sum of these products of deviations from the means is the basis of the **correlation coefficient**. The only issue with this technique is that the size of the correlation coefficient depends on the units of measure of the variables.

To circumvent this problem, we take the sum of the products of the deviations from the mean, and divide by the square root of the product of the sum of the squares of the deviations. In doing so, the coefficient becomes independent of the variable's measurement scales. Using one of the above data sets, you can confirm that this does the trick and is independent of the units of measure!

Definition

The observed value r of the sample linear correlation coefficient is

defined as $r = \dfrac{\sum\limits_{i=1}^{n}(x_i - \bar{x})(y_i - \bar{y})}{\sqrt{\sum\limits_{i=1}^{n}(x_i - \bar{x})^2 \sum\limits_{i=1}^{n}(y_i - \bar{y})^2}} = \dfrac{\sum d_x d_y}{\sqrt{\sum d_x^2 \sum d_y^2}}$

Values of r near 0 indicate a weak correlation between X and Y, and values of r nearer ± 1 indicate a strong positive or strong negative correlation.

Correlation and causation

It is important at this point to discuss the distinction between correlation and causation of two random variables, X and Y. Correlation between two variables appears to imply dependence, i.e. as one variable changes, so does the other. However, we must be wary not to interpret the correlation between two variables as causation, i.e. a change in one variable *causes* a change in the other. If there appears to be causation, this might be mere coincidence, or other factors could be at work. Finding correlation is the job of the statistician, but asserting causation is the job of the specialist in the respective field of study. This will be further discussed later in the chapter.

? On January 28 1986, millions of people watched the Space Shuttle *Challenger* break apart after just 73 seconds into the flight. All seven crew members died, among them Christa McAuliffe, the first teacher ever to be invited to be part of an astronaut team. The cause of the disaster was the failure of an O-ring on the right solid rocket booster that prevents hot gases escaping from the rockets (O-rings help seal the joints of different segments of the solid rocket boosters). It is now known that a leading factor in the O-ring failure was the exceptionally low temperature (about 31°F) at the time of the launch. After this disaster, the strength of the relationship, or the correlation, between temperature at the time of the launch and O-ring erosion was found to be of fundamental importance during the launch phase of a space shuttle.

In Example 1, we will use a GDC to calculate the linear correlation coefficient, r.

Example 1

The following table shows the prevalence of smoking, as a percent, in some states of the USA and the incidence of smoking-attributable deaths.

State	X- Prevalence of Smoking	Y- Smoking attributable death rate
Alaska	42	34
California	2	18
D.C.	26	14
Florida	22	20
Georgia	15	41
Indiana	44	44
Kentucky	50	51
Missouri	38	45
New York	12	17
Rhode I.	28	27
Texas	22	28
Utah	1	1

Data taken from: http://www.cdc.gov/tobacco/data_statistics/state_data/data_highlights/2006/pdfs/datahighlights06table5.pdf

a Draw a scatter diagram to illustrate this data. Include the mid point, and draw an axis with origin at the mean point.
b Discuss the correlation between X and Y.
c Use a GDC to calculate r.

---

a

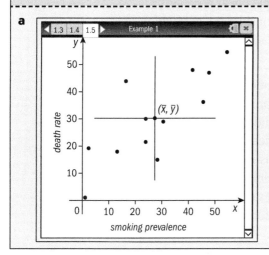

Find $(\bar{x}, \bar{y})$ and use the GDC to plot the points, including the mean point.

b There is a positive correlation between X and Y since most of the points are in the 1st and 3rd quadrants.

Determine the quadrants where most points lie after drawing axes through the mean point.

Use a GDC to calculate the linear correlation coefficient, r.

c

	A xc...	B yc...	C	D
=				=LinRegMx('xcoord,'y
2	2	18	Reg...	m*x ≠ b
3	26	14	m	0.7377
4	22	20	b	9.7671
5	15	41	r^2	0.62231
6	44	44	r	0.78887

A6 44

From the GDC, $r = 0.789$, hence the variables have a positive correlation.

In the examples so far, we have considered the relationship between two variables using their observed, or experimental, values. If we are, however, to make inferences about the correlation between measurements in an entire population (in order to create mathematical models), then we need to define what is meant by the correlation of two **random** variables, X and Y.

Sampling distributions

We have considered r, the observed value of the sample linear correlation coefficient, using n paired observations. In order to consider the correlation of the population, we must analyze the distribution of each r, where each sample will probably give a different r-value.

Since there are many possible samples x_i, the distribution of random variables X_i would be the same as the distribution of the random variable X, and similarly the distribution of random variables Y_i would be the same as the distribution of the random variable Y. The sample means, $\bar{x}$ and $\bar{y}$, calculated for all possible sample sizes will follow a sampling distribution, $\bar{X}$ and $\bar{Y}$. In the same way, all values of r calculated from all possible samples will form a sampling correlation coefficient distribution R. Each sample correlation coefficient can be used as an estimate of the population correlation coefficient ρ (rho).

The **sample product moment correlation coefficient R**, for n paired observations (x, y) on X and Y, is

$$R = \frac{\sum_{i=1}^{n}(X_i - \bar{X})(Y_i - \bar{Y})}{\sqrt{\sum_{i=1}^{n}(X_i - \bar{X})^2 \sum_{i=1}^{n}(Y_i - \bar{Y})^2}}$$

If we take the numerator of R, we can multiple out the brackets to obtain $\sum_{i=1}^{n}(X_i - \bar{X})(Y_i - \bar{Y}) = \sum_{i=1}^{n}(X_iY_i - \bar{X}Y_i - X_i\bar{Y} + \bar{X}\bar{Y})$

Since $\bar{X}$ and $\bar{Y}$ are constants, this expression is equal to

$$\sum_{i=1}^{n}(X_iY_i) - \bar{X}\sum_{i=1}^{n}Y_i - \bar{Y}\sum_{i=1}^{n}X_i + \sum_{i=1}^{n}\bar{X}\bar{Y}$$

Since $\bar{X} = \dfrac{\sum_{i=1}^{n}X_i}{n}, \bar{Y} = \dfrac{\sum_{i=1}^{n}Y_i}{n}$, therefore,

$$\sum_{i=1}^{n}(X_iY_i) - \bar{X}\sum_{i=1}^{n}Y_i - \bar{Y}\sum_{i=1}^{n}X_i + \sum_{i=1}^{n}\bar{X}\bar{Y} = \sum_{i=1}^{n}(X_iY_i) - 2n\bar{X}\bar{Y} + n\bar{X}\bar{Y}$$

$$= \sum_{i=1}^{n}(X_iY_i) - n\bar{X}\bar{Y}$$

Hence, $\sum_{i=1}^{n}(X_i - \bar{X})(Y_i - \bar{Y}) = \sum_{i=1}^{n}(X_iY_i) - n\bar{X}\bar{Y}$, and an alternative

formula for R is therefore $R = \dfrac{\sum_{i=1}^{n}X_iY_i - n\bar{X}\bar{Y}}{\sqrt{\sum_{i=1}^{n}(X_i^2 - n\bar{X}^2)(Y_i^2 - n\bar{Y}^2)}}$

We can classify the strength of the correlation between the random variables X and Y by using the following *general* classification:

- ± 1 indicates a perfect positive/negative correlation.
- $0.5 \leq R < 1$ indicates a strong positive correlation.
- $-1 \leq R \leq -0.5$ indicates a strong negative correlation.
- $0.1 \leq R < 0.5$ indicates a weak positive correlation.
- $-0.5 < R \leq -0.1$ indicates a weak negative correlation.
- $-0.1 < R < 0.1$ indicates a highly weak correlation, or no correlation.

Of course, if $R = 0$, X and Y may not have a linear correlation, but they could have a different correlation, e.g. quadratic, exponential, sinusoidal, etc. These kinds of correlation, however, are not part of this course.

As stated earlier for values of r, the values of R near 0 indicate a weak correlation between X and Y, and values of R nearer ± 1 indicate a strong positive or negative correlation.

Example 2

Using the table below, draw a scatter diagram, and use one of the formulae above to calculate the product moment correlation coefficient, R, between the two random variables X (percentage of economic growth rate) and Y (percentage of Standard and Poor's 500 returns rate). Check this result by using a GDC to find r, and interpret your result.

 The Standard and Poor's 500 index is a stock market index based on the total value of issued shares of stock of 500 publicly traded companies. The S&P 500 returns rate is the percentage you would have earned had you invested your money in all 500 companies.

$X\%$ of economic growth	$Y\%$ of S&P 500 returns
1.9	7.7
2.6	12.6
3.9	13.7
3.2	9.8

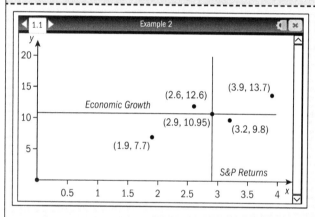

Make sure that the mean point is included.

Method I

$$\sum_{i=1}^{n}[(x-\bar{x})(y-\bar{y})] = (1.9-2.9)(7.7-10.95)+(2.6-2.9)(12.6-10.95)$$

$$+ (3.9-2.9)(13.7-10.95)+(3.2-2.9)(9.8-10.95) = 5.16$$

$$\sum_{i=1}^{n}(x-\bar{x})^2 = (1.9-2.9)^2+(2.6-2.9)^2+(3.9-2.9)^2+(3.2-2.9)^2 = 2.18$$

$$\sum_{i=1}^{n}(y-\bar{y})^2 = (7.7-10.95)^2+(12.6-10.95)^2$$

$$+ (13.7-10.95)^2+(9.8-10.95)^2 = 22.17$$

$$\therefore r = \frac{5.16}{\sqrt{2.18\times22.17}} \approx 0.742$$

Using the formula for r, r = 0.742.

Method II

◆	A xcoord	B ycoord	C	D	E
=				= TwoVar(	
2	2.6	12.6	$\bar{x}$	2.9	
3	3.9	13.7	Σx	11.6	
4	3.2	9.8	Σx^2	35.82	
5			sx : = sn−...	0.852447	
6			σx : = σn...	0.738241	
7			n	4.	
8			$\bar{y}$	10.95	
9			Σy	43.8	
10			Σy^2	501.78	
11			sy : = sn−...	2.71846	
12			σy : = σn...	2.35425	
13			Σxy	132.18	
14			r	0.74223	

D26 = 22.17

$r = 0.742$

There is a strong positive correlation between the two random variables.

(You will notice that the GDC gives you all the values necessary to evaluate r manually.)

Interpret the r value.

Although it was the French physicist Auguste Bravais who first developed the correlation coefficient in the mid 1800s, it was the English mathematician Karl Pearson (1857–1936) who formalized this concept. He borrowed the concept of 'moments' from Physics, hence the name the 'Pearson Product Moment Correlation Coefficient'. It is the most common method of computing a correlation coefficient between two variables that are linearly related. Pearson founded the world's first university statistics department at University College London in 1911. Either r or R can be used for the Pearson sample product moment correlation coefficient.

Any correlation coefficient must be treated with great care. For example, in the UK in the early 1930s there was seen to be a statistically significant relationship between an increase in radio licenses and an increase in mental illness. It would be absurd to assume that one is the cause of the other; there are probably many other unrelated factors at work here, and so we call this a *spurious correlation*. The data may be affected by a *third factor, or more factors*, to which both X and Y are related. For example, there could be a correlation between smoking and high blood pressure, however they could both be related to a third variable (stress level) which could actually be causing both the high blood pressure and the desire to smoke.

Hence, great caution should always be taken before asserting that a change in one variable causes a change in the other. Many other factors should be considered, and other statistical tests performed. This is why, as stated earlier, finding a correlation is the job of the statistician, but asserting causation should be left to the specialists in their respective fields of study.

Exercise 4A

For the data sets below, draw a scatter diagram, including the mean point and axis through the point. Calculate R for the random variables X and Y by using a GDC, and interpret this value.

1 The opinions of ten people were sought regarding their levels of satisfaction with how a daily newspaper (X), and a television news report (Y), informed them of world affairs. Their levels of satisfaction are shown in the table below, with 1 being the lowest and 5 being the highest satisfaction.

X	4	1	3	2	1	2	5	4	4	2
Y	1	1	2	4	3	4	2	1	1	1

2 The lengths in mm (X) and widths in mm (Y) of certain leaves are given in the table below.

X	100	115	140	149	88	132	152	144	121	128
Y	33	38	40	51	36	40	51	43	32	42

3 The percentages scored in the same subject on two different tests are shown below. Let X be the percentage scored in Test 1, and Y be the percentage scored in Test 2.

X	55	35	66	82	91	79	48	52	71	88
Y	58	65	52	35	36	42	60	55	50	38

4.2 Covariance

Another statistical measure used to determine if there is a relationship between two random variables is called the covariance. As the name implies, the covariance tells us whether or not the variables *vary* together.

- A positive covariance indicates that *higher than average* values of one variable tend to be paired with *higher than average* values of the other variable.
- A negative covariance indicates that *higher than average* values of one variable tend to be paired with *lower than average* values of the other variable.
- Zero covariance indicates that there is no correlation between the variance of each variable.

In other words, we find the sum of the products of the individual values' deviations from the mean, i.e. the numerator of r, and take the average of this sum. This is the equivalent of dividing the numerator of the formula for r by n and we thus obtain the sample covariance:

$$\text{Cov}(X, Y) = \frac{\sum_{i=1}^{n}[(X_i - \bar{X})(Y_i - \bar{Y})]}{n} \quad \text{or} \quad \text{Cov}(X, Y) = \frac{\sum_{i=1}^{n}XY}{n} - \bar{X}\bar{Y}$$

The covariance, however, is not dimensionless, i.e. it is dependent on the units of the data. This means a value that represents a strong relationship in one unit, might not indicate the same relationship if a different unit is used. For this reason, when we have a distribution of two random variables X and Y, the correlation coefficient r addresses this issue by normalizing the covariance to the product of the standard deviations of the variables, creating a quantity between -1 and $+1$ that is independent of units. This facilitates the comparison of different data sets.

The covariance indicates whether there is a positive or negative relationship, but it does not tell us anything about the strength of the relationship.

If we use the first definition given for R,

i.e. $R = \dfrac{\sum_{i=1}^{n}(X_i - \bar{X})(Y_i - \bar{Y})}{\sqrt{\left(\sum_{i=1}^{n}X_i^2 - n\bar{X}^2\right)\left(\sum_{i=1}^{n}Y_i^2 - n\bar{Y}^2\right)}}$

and divide the numerator and denominator by n and replace $\bar{X}$ by $E(X)$ and replace $\bar{Y}$ by $E(Y)$, the sums become averages.

Replacing these averages by the expected values, we obtain,

$$\rho = \frac{E[(X - E[X])(Y - E[Y])]}{\sqrt{E[(X - E[X])^2]E[(Y - E[Y])^2]}}$$

The denominator is equal to $\sqrt{\text{Var}(X)\text{Var}(Y)}$. The numerator looks like a variance, except it contains a product of deviation of X and Y, rather than a square of a deviation, and in fact is called the covariance of X and Y, Cov(X, Y), and the greek letter rho, ρ, is the population product moment coefficient.

Hence, $\rho = \dfrac{\text{Cov}(X,Y)}{\sqrt{\text{Var}(X)\text{Var}(Y)}}$

Definition: For each set of n paired observations, the covariance of the random variables X and Y of a joint distribution, Cov(X, Y), is a measure of the mean value of the product of the two variables' deviations from their respective means, i.e.

$$\text{Cov}(X, Y) = E[(X - E(X))(Y - E(Y))] = E[(X - \mu_x)(Y - \mu_y)],$$

where $\mu_x = E(X)$, $\mu_y = E(Y)$

Starting with this definition we can derive an alternative one.

$$\begin{aligned}
\text{Cov}(X, Y) &= E[(X - \mu_x)(Y - \mu_y)] \\
&= E[XY - X\mu_y - Y\mu_x + \mu_x\mu_y] \\
&= E(XY) - \mu_y E(X) - \mu_x E(Y) + \mu_x\mu_y \\
&= E(XY) - \mu_y\mu_x - \mu_x\mu_y + \mu_x\mu_y \\
&= E(XY) - \mu_x\mu_y
\end{aligned}$$

Properties of covariance

1 Symmetry: Cov(X, Y) = Cov(Y, X), and Var(X) = Cov(X, X)

The proof of this is left as an exercise.

2 For two independent events X and Y, Cov(X, Y) = 0.

Proof: If X and Y are independent variables, then

$E(XY) = E(X)E(Y) = \mu_x\mu_y$ and $\text{Cov}(X, Y) = E(XY) - \mu_x\mu_y$
$\qquad = \mu_x\mu_y - \mu_x\mu_y = 0.$

The converse, however, is not necessarily true: if Cov(X, Y) = 0, it does not follow that X and Y are independent. For example, let U be the random variable 'the number appearing on the first throw of a die', and let V be the random variable 'the number appearing on the second throw of a die'. If $X = U+V$ and $Y = U-V$, then $E(XY) = E(X)E(Y)$; hence, Cov(X, Y) = 0, however the random variables X and Y are not independent.

Students should verify the above result as an exercise.

From the above properties, we can establish the following theorems:

Theorem 1: If X and Y are independent variables, then $\rho = 0$.

The proof is left as an exercise.

Theorem 2: If X and Y have a linear relationship, i.e. $Y = mX + c$, then $\rho = \pm 1$.

Proof:

$Y = mX + c \Rightarrow E(Y) = mE(X) + c$. Hence,

$\mu_y = m\mu_x + c$; $\text{Var}(Y) = m^2 \text{Var}(X)$.

Therefore,

$E(XY) = E(mX^2 + cX) = mE(X^2) + c\mu_x$

and, using the formula for ρ and making the substitutions above,

$$\rho = \frac{mE(X^2) + c\mu_x - \mu_x(m\mu_x + c)}{\sqrt{\text{Var}(X)m^2\text{Var}(X)}} = \frac{mE(X^2) + c\mu_x - (m\mu_x^2 - u_x c)}{\sqrt{m^2(\text{Var}(X))^2}}$$

$$= \frac{m(E(X^2) - \mu_x^2)}{|m|\text{Var}(X)} = \frac{m\text{Var}(X)}{|m|\text{Var}(X)} = \frac{m}{|m|} = \pm 1$$

Practically speaking, we can estimate ρ for the population only by using the correlation coefficient of a sample drawn from the population. Hence, an estimate for ρ is R, the sample product moment coefficient.

Example 3

Given that $X \sim N(0, 1)$ and $Y = 2X$, show that the covariance of the joint distribution X and Y, $\text{Cov}(X, Y)$, is 2.

$\text{Cov}(X, Y) = E(XY) - E(X)E(Y) = E(2X^2) - E(X)E(2X)$	*Use definition.*
$E(2X^2) = 2E(X^2)$	*Use properties of expectation algebra.*
$\text{Var}(X) = E(X^2) - [E(X)]^2 \Rightarrow E(X^2) = \text{Var}(X) + [E(X)]^2$	
Hence,	
$E(X^2) = 1 + 0 = 1$	
Hence, $\text{Cov}(X, Y) = E(2X^2) - E(X)E(2X)$	*Substitute $E(X) = 0$ and $\text{Var}(X) = 1$.*
$\qquad = 2E(X^2) - E(X)2E(X) = 2 - 0 = 2$	*Evaluate.*

Exercise 4B

Prove the following:

1 $\text{Cov}(X, Y) = \text{Cov}(Y, X)$
2 $\text{Cov}(X, X) = \text{Var}(X)$
3 $\text{Cov}(aX, Y) = a\text{Cov}(X, Y)$
4 $\text{Cov}(X, bY) = b\text{Cov}(X, Y)$
5 $\text{Cov}(X_1 + X_2, Y) = \text{Cov}(X_1, Y) + \text{Cov}(X_2, Y)$
6 $\text{Cov}(X, Y_1 + Y_2) = \text{Cov}(X, Y_1) + \text{Cov}(X, Y_2)$
7 $\text{Var}(X + Y) = \text{Var}(X) + \text{Var}(Y) + 2\text{Cov}(X, Y)$
8 If X and Y are independent variables, then $\rho = 0$.

4.3 Hypothesis testing

Introduction

This introduction serves to aid understanding of the bivariate normal distribution, but will not be formally assessed.

Before we can determine how large the correlation coefficient must be in order to conclude that there is a significant correlation between two random variables X and Y, we will perform an appropriate sample statistic. We will first, however, define what we mean by **bivariate normal distribution**.

You are already very familiar with the normal distribution for a single continuous random variable. We will now consider the joint normal distribution for two random variables X and Y.

Let X and Y be two normally distributed variables, i.e.

$$X \sim N(\mu_X, \sigma_X^2) \text{ and } Y \sim N(\mu_Y, \sigma_Y^2).$$

Then, the bivariate normal distribution, or the joint normal distribution, is defined by the probability density function:

$$f(x,y) = \frac{1}{2\pi\sigma_X\sigma_Y\sqrt{1-\rho^2}} e^{\left[-\frac{1}{2(1-\rho^2)}\left[\left(\frac{x-\mu_x}{\sigma_x}\right)^2 + \left(\frac{y-\mu_y}{\sigma_y}\right)^2 - 2\rho\left(\frac{x-\mu_x}{\sigma_x}\right)\left(\frac{y-\mu_y}{\sigma_y}\right)\right]\right]}$$

For standardized variables, i.e. $z_x = \dfrac{x-\mu_x}{\sigma_x}; z_y = \dfrac{y-\mu_y}{\sigma_y}$, the bivariate

normal PDF becomes $f(z_x z_y) = \dfrac{1}{2\pi\sqrt{1-\rho^2}} e^{-\frac{z_x^2+z_y^2-2\rho z_x z_y}{2(1-\rho)^2}}$

You know that the shape of the normal distribution of a single random variable is that of a 2D bell-shaped surface. The bivariate normal distribution has a 3D bell-shaped surface.

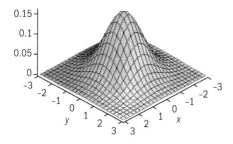

You have already seen that if X and Y are two independent random variables, then $\text{Cov}(X, Y) = 0$ and hence $\rho = 0$. For X and Y having a bivariate normal distribution, the converse is also true.

Furthermore, we can now state the following.

Theorem 3

For X and Y with a bivariate normal distribution, X and Y are independent if and only if $\rho = 0$.

The first part, i.e. $\Rightarrow$, was already seen in Theorem 1.
For $\Leftarrow$, we have the definition of a joint probability density function

of X and Y, i.e. $f(x,y) = \dfrac{1}{2\pi\sigma_X\sigma_Y\sqrt{1-\rho^2}}\, e^{\left[-\frac{1}{2(1-\rho^2)}\left[\left(\frac{x-\mu_X}{\sigma_x}\right)^2+\left(\frac{y-\mu_y}{\sigma_y}\right)^2-2\rho\left(\frac{x-\mu_X}{\sigma_x}\right)\left(\frac{y-\mu_y}{\sigma_y}\right)\right]\right]}$

Using $\rho = 0$ and the laws of exponents,

$$f(x,\ y) = \frac{1}{2\pi\sigma_X\sigma_Y}\, e^{-\frac{1}{2}\left(\frac{x-\mu_X}{\sigma_x}\right)^2} \times \frac{1}{2\pi\sigma_X\sigma_Y}\, e^{-\frac{1}{2}\left(\frac{y-\mu_y}{\sigma_y}\right)^2} = f(x) \times f(y)$$

Hence, any probabilities calculated from the joint distribution of X and Y will result in $P(x_1 \le X \le x_2,\ y_1 \le Y \le y_2) = P(x_1 \le X \le x_2) \times P(y_1 \le Y \le y_2)$, and therefore X and Y are independent.

t-Statistic for dependence of *X* and *Y*

For a bivariate normal distribution of X and Y, the correlation coefficient can be used to determine, at a given level of significance, whether X and Y have a linear correlation by determining if $\rho = 0$. We will use Theorem 4 to make this determination.

Theorem 4

If X and Y have a bivariate normal distribution such that $\rho = 0$, then the sampling distribution $R\sqrt{\dfrac{n-2}{1-R^2}}$ has the *student's t-distribution* with $(n-2)$ degrees of freedom.

This means that for any random sample of n independent paired data from the joint normal distribution $(X,\ Y)$ with $\rho = 0$, the observed value r of the sample product moment coefficient R has the property that $t = r\sqrt{\dfrac{n-2}{1-r^2}}$, i.e. it is distributed as *student's t-statistic* with degrees of freedom $v = n - 2$.

To perform a hypothesis test with $H_0: \rho = 0$ and $H_1: \rho \neq 0$ on a random sample of n independent pairs $(x_1,\ y_1),\ (x_2,\ y_2),\ \ldots\ (x_n,\ y_n)$ from a bivariate normal distribution $(X,\ Y)$ we:

- calculate r, the observed value of the sample product moment correlation coefficient R
- calculate the *t*-statistic, $t = r\sqrt{\dfrac{n-2}{1-r^2}}$
- calculate the critical values for the indicated level of significance, or calculate the *p*-value.

Example 4

Test, at the 10% significance level, whether the data from Example 2 shows significant evidence of correlation between the random variables X (percentage of economic growth rate) and Y (percentage of Standard and Poor's 500 returns rate).

$H_0: \rho = 0$; $H_1: \rho \neq 0$	*Step 1: State the null and alternative hypotheses.*
$t = r\sqrt{\dfrac{n-2}{1-r^2}} \Rightarrow t = 0.742\sqrt{\dfrac{4-2}{1-0.742^2}} \approx 1.565$	*Step 2: Find the value of the test statistic.*
For H_0, $T : t_{(2)}$	
$p = P(T \leq -1.565) + P(T \geq 1.565) = 0.258$	

Step 3: Using a GDC, calculate the p-value.

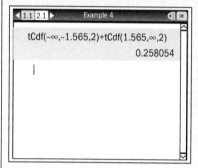

tCdf(-∞,-1.565,2)+tCdf(1.565,∞,2)
  0.258054

Alternatively, using the 'Linear Regression t test' in the Stats tests menu of the GDC, we obtain the following:	*Confirm answers above using the GDC.*

	A xdots	B ydots	C	D	E
=					=LinRegtT
1	1.9	7.7		Title	Linear R...
2	2.6	12.6		Alternate...	β & ρ ≠...
3	3.9	13.7		RegEqn	a+b*x
4	3.2	9.8		t	1.56634
5				PVal	0.25777
6				df	2.
7				a	4.08578
8				b	2.36697
9				s	2.23119
10				SESlope...	1.51115
11				r^2	0.550906
12					0.74222

E1 ="Linear Reg t Test"

Since 0.258 > 10%, we do not have enough evidence to reject H_0. Hence, there is not evidence of significant correlation between the random variables X and Y.	*Step 4: Reading the p-value from the results, we compare with the significance level and state conclusion.*

Note: Although the r value seems relatively high, with so few data values the correlation is not significant.

Exercise 4C

1 Use the data in Example 1 to determine at the 1% significance level whether the random variables show a significant level of correlation.

2 Below is a table of fathers' and sons' heights, in inches. Determine, at the 5% significance level, if there is significant evidence of correlation between the random variables X and Y.

X-Father's height	64	66	68	69	70	72	74
Y-Son's height	69	64	67	64	62	73	71

3 Using the data below, determine, at the 10% significance level, if there is significant evidence of correlation between the random variables X (height of a supermodel, in inches) and Y (weight of a supermodel, in pounds).

Height	67	70	70	70	71	72	72	72	73
Weight	104	116	121	126	117	113	124	126	127

4.4 Linear regression

At the start of this chapter we graphed pairs of data sets and determined through the scatter diagrams whether there was a correlation between the two variables, and if so, whether the correlation was positive or negative. If the scatter diagrams showed a correlation between the two variables, we could then draw a line of best fit, or a line of regression.

The regression line of Y on X, $E(Y)|X = x$, is used to estimate Y given that the values of X are accurate. In other words, the linear regression of Y on X focuses on the conditional distribution of Y given X, rather than on the joint distribution of Y and X. For example, X can be the price of a particular stock, and Y might be the amount of stock sold. In this case, the value of the stock is known, but the number of shares bought will vary according to the price. Hence, X is the independent variable, and Y is the dependent variable.

One method of calculating the regression line is to find the least sum of the squares of the vertical distances from the points, i.e. the area of the squares of the distances is minimized, as the figure below illustrates.

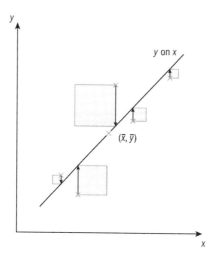

On YouTube you can watch Douglas Butler, creator of the software Autograph, demonstrate finding the least squares regression line of Y on X.

The sum of the squares of the vertical distances from each point to the line is obtained, and the line producing the least sum is the regression line of Y on X.

Using the principle of least squares, the Y on X regression line is given by the following formulae:

$$y - \bar{y} = \left(\frac{\sum_{i=1}^{n} [(x_i - \bar{x})(y_i - \bar{y})]}{\sum_{i=1}^{n} (x_i - \bar{x})^2} \right)(x - \bar{x}) \text{ or } y - \bar{y} = \left(\frac{\sum_{i=1}^{n} x_i y_i - n\bar{x}\,\bar{y}}{\sum_{i=1}^{n} x_i^2 - n\bar{x}^2} \right)(x - \bar{x})$$

The derivation of these formulae is beyond the level of this course.

Notice that the gradient of the Y on X regression line is the quotient of $\text{Cov}(X, Y)$ and $\text{Var}(x)$.

Hence, if the scatter diagrams show a correlation between the two variables, then we can draw in a line of best fit, or line of regression, of Y on X, $E(Y)|X = x$. This is used to estimate Y given that the values of X are accurate.

The mathematician Francis Galton first discussed the idea of linear regression. In a famous study, he measured the mean height of parents, and the mean height of their children, both in inches. The following table includes some values from his initial research.

mean height of parents (X)	72.5	70.5	68.5	66.5	64.5
mean height of children (Y)	71.2	69.5	68.2	67.2	65.8

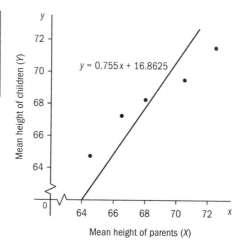

From the entire table of values and regression, Galton noticed that on average, the heights of adults with tall parents were higher than the heights of adults with short parents. More interestingly, he noticed that on average, adult offspring of tall parents are shorter than their parents, whereas adult offspring of short parents are taller than their parents.

For Galton's full table of values and regression see http://www.math.uah.edu/stat/data/Galton.html

Using a GDC to calculate r, we obtain that $r = 0.981$, indicating a strong positive correlation between the variables X and Y.

	A xcoord	B ycoord	C	D	E
=					=TwoVar(
2	70.5	69.5		x	68.5
3	68.5	68.2		Σx	342.5
4	66.5	67.2		Σx²	23501.3
5	64.5	65.8		sx := sn−...	3.16228
6				σx := σn...	2.82843
7				n	5.
8				y	68.58
9				Σy	342.9
10				Σy²	23539.8
11				sy := sn−...	2.43557
12				σy := σn...	2.17844
13				Σxy	23518.9
14				r	0.980272

C

It is important to state here that the line of best fit, or line of linear regression, can only be used to estimate unknown values contained within the interval of the data (interpolation). For estimating any values outside this interval (extrapolation), an assumption that the pattern continues is not a valid assumption.

As you have already seen, there are different ways to write the formulae for Cov(X, Y), Var(X), and Var(Y). Depending on the kind of information you are given to work with, some formulae will be more appropriate than others.

Instead of being given the data directly, we may be given summary statistics and asked to find the line of regression. In this case, we must see which formula fits the given statistics best. For example, if we are given the following summary statistics:

$$n = 10; \sum xy = 906.82; \sum x = 103.8; \sum y = 100.9; \sum x^2 = 1983.08; \sum y^2 = 1050.67$$

and asked to find the regression line of Y on X for these values, the

formula best suited is $y - \bar{y} = \left(\dfrac{\sum_{i=1}^{n} x_i y_i - n\bar{x}\,\bar{y}}{\sum_{i=1}^{n} x_i^2 - n\bar{x}^2} \right)(x - \bar{x})$

Chapter 4   143

Using the given values, the gradient of the line of regression is:

$$m = \dfrac{\left(\displaystyle\sum_{i=1}^{n} x_i y_i - n\bar{x}\,\bar{y}\right)}{\displaystyle\sum_{i=1}^{n} x_i^2 - n\bar{x}^2} = \dfrac{906.82 - 10 \times \dfrac{103.8}{10} \times \dfrac{100.9}{10}}{1983.08 - 10 \times \left(\dfrac{103.8}{10}\right)^2} = -0.1551\ldots$$

Hence, $y - 10.09 = -0.155\,(x - 10.38) \Rightarrow y = -0.155x + 11.7$.

Most of the time, however, we should have the data to work with directly, as the following example shows.

Example 5

The data below shows the fuel consumption and average speed of a typical Euro 4 passenger car when driven on a test circuit.

X-Average speed in km/h	10	15	20	30	35	40	45	50
Y-Fuel consumption in litres/100 km	13	10	9	7	6.6	6.2	6	6

a Justify that it makes sense to find the line of best fit, or linear regression, for the variables in the table.

b Use the data to find the equation of the linear regression, and check your answer on a GDC.

c Use the equation to estimate the fuel consumption of a typical Euro 4 passenger car travelling at an average speed of 25 km/h.

d Does it make sense to use the table above to find the fuel consumption of a car traveling at 80 km/h?

e Interpret your answers for this table of values.

a

	A xvalues	B yvalues	C	D	E
=					=TwoVar(
3	20	9		Σx	245.
4	30	7		Σx²	8975.
5	35	6.6		sx := sn-...	14.5006
6	40	6.2		σx := σn...	13.5641
7	45	6		n	8.
8	50	6		ȳ	7.975
9				Σy	63.8
10				Σy²	553.
11				sy := sn-...	2.51268
12				σy := σn...	2.3504
13				Σxy	1719.
14				r	-0.9209...

E1	="Two–Variable Statistics"

Since $r \approx -0.921$, this indicates a strong negative correlation, hence we can draw a line of best fit.

b From the GDC above, we obtain the following values:

$$\sum xy = 1719; \quad \bar{x} = 30.625; \quad \bar{y} = 7.975$$

$$n = 8; \sum x_i^2 = 8975. \text{ Hence,}$$

$$m = \frac{\sum x_i y_i - n\bar{x}\,\bar{y}}{\sum x_i^2 - n\bar{x}^2} = \frac{1719 - 8 \times 30.625 \times 7.975}{8975 - 8 \times 30.625^2} \approx -0.1595\ldots$$

$$y - \bar{y} = \frac{\sum x_i y_i - n\bar{x}\,\bar{y}}{\sum x_i^2 - n\bar{x}^2}(x - \bar{x}) \Rightarrow y - 7.975 = -0.1595(x - 30.625)$$

Hence, $y = -0.160x + 12.9$ (to 3 s.f.)

#	A xvalues	B yvalues	C	D	E
=					=LinRegMx('xvalues, 'yvalues,1
1	10	13	Title	Linear Regression (mx+b)	
2	15	10	RegEqn	m*x+b	
3	20	9	m		-0.159575
4	30	7	b		12.862
5	35	6.6	r²		0.848066
6	40	6.2	r		-0.920905
7	45	6	Resid	{1.7337579617839, -0.46836…	
8	50	6			
9					
10					
11					
12					
B12					

From the GDC, we see that the regression line is $y = -0.160x + 12.9$.

c $y = -0.1596(25) + 12.86 = 8.87$, hence the fuel consumption at a constant speed of $25 \,\mathrm{km/h}$ would be about 8.9 litres/100 km.

Use the equation to substitute for x and find y.

d No, since 80 lies outside the data domain.

Check if the value lies within the data domain.

e Since these speeds are averages, they indicate driving in urban or congested areas where there is more traffic, traffic lights, and congestion in general. The car is therefore in a continual 'stop-start' mode, which causes increased fuel usage.

Attempt a reasonable explanation, assuming that the data is correct.

Example 6

The table below continues for the following speeds of the same test cars on the same test circuit.

X-Average speed in km/h	50	55	60	65	70	75	80
Y-Fuel consumption in litres/100 km	6	6	6	6	6	6.1	6.1

Describe and interpret the relationship between the two variables.

The line joining the points is just about parallel to the x-axis, hence its gradient is 0. Driving at speeds allowed on priority roads, i.e. between 50 and 80 km/h, the fuel consumption is almost constant, 6 litres/100 km.

Example 7

The table below continues for the following speeds of the same test cars on the same test circuit.

X-Average speed in km/h	80	85	90	95	105	110	115	120
Y-Fuel consumption in litres/100 km	6.1	6.3	6.4	6.4	6.6	6.7	6.8	6.9

a Justify that it makes sense to find the line of best fit, or linear regression, for the variables in the table.

b Use a GDC to find the equation of the linear regression line.

c Use the equation to estimate the fuel consumption of a car travelling at an average speed of 100 km/h.

d Interpret your answers for this table of values.

e Determine, at the 10% significance level, if there is a correlation between the average speeds and fuel consumption of the test cars for average speeds between 10 and 50 km/h, and between 80 and 120 km/h.

---

a

Use a GDC to find r.

Since $r \approx 0.989$, this indicates a strong positive correlation, hence we can find the regression line.

b

Use the GDC to find the line of best fit.

The equation of the line of best fit is:
$y = 0.0183x + 4.69$ (to 3 s.f.)

c $y = 0.01833(100) + 4.692 = 6.53$, hence the fuel consumption at 100 km/h is about 6.5 litres/100 km

If the value is in the interval of the data, then substitute for x and evaluate y.

Attempt a reasonable explanation, assuming that the data is correct.

d The speeds in the table indicate highway traffic with little interruption, and so it makes sense that as speeds increase so will fuel consumption.

e For the data $10 \le X \le 50$:

$H_0: \rho = 0$; $H_1: \rho \ne 0$

$$t = r\sqrt{\frac{n-2}{1-r^2}} \Rightarrow t = -0.921\sqrt{\frac{8-2}{1-0.921^2}} = -5.79106$$

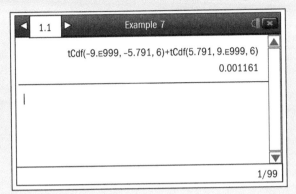

$p = P(T \le -5.791) + P(T \ge 5.791) = 0.00116$

$0.00116 < 0.1$, hence we reject H_0, since there is significant evidence of correlation between the random variables X and Y.

For the data $80 \le X \le 120$:

$H_0: \rho = 0$; $H_1: \rho \ne 0$

$$t = r\sqrt{\frac{n-2}{1-r^2}} \Rightarrow 0.989426\sqrt{\frac{8-2}{1-0.989426^2}} = 16.71$$

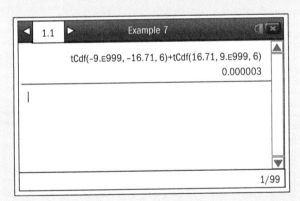

$p = P(T \le -16.71) + P(T \ge 16.71) = 0.000003$

$0.000003 < 0.1$, hence we reject H_0, since there is significant evidence of correlation between the random variables X and Y.

Step 1: State the null and alternative hypothesis.

Step 2: Find t.

Step 3: Find the p-value and confirm using the stats test menu of the GDC.

Step 4: State conclusion.

Step 1: State the null and alternative hypotheses.

Step 2: Find t.

Step 3: Find the p-value, and confirm using the stats test menu of the GDC.

Step 4: State conclusion.

To calculate the regression line of X on Y, we find the least sum of the squares of the horizontal distances from the points, i.e. the area of the squares of the distances is minimized.

In other words, the linear regression of X on Y focuses on the conditional distribution of X given Y, rather than on the joint probability distribution of X given Y.

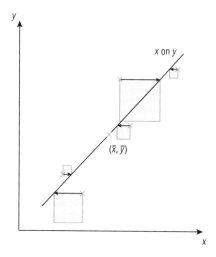

The formula for the X on Y regression line is

$$x - \bar{x} = \left(\frac{\sum_{i=1}^{n}(x_i - \bar{x})(y_i - \bar{y})}{\sum_{i=1}^{n} y_i - \bar{y}^2} \right)(y - \bar{y}) = \left(\frac{\sum_{i=1}^{n} x_i y_i - n\bar{x} \cdot \bar{y}}{\sum_{i=1}^{n} y_i^2 - n\bar{y}^2} \right)(y - \bar{y})$$

Notice that the gradient of the X on Y regression line is the quotient of $\mathrm{Cov}(X, Y)$ and $\mathrm{Var}(Y)$.

It is important to remember that both the Y on X regression line and the X on Y regression line pass through the mean point $(\bar{x}, \bar{y})$, as discussed in the beginning of this chapter in drawing scatter diagrams.

Example 8

A teacher gave two different Statistics tests to his class with the following paired results, given in percentages.

X-Test1	65	88	83	92	50	67	100	100	73	90	83	94
Y-Test2	52	57	78	76	30	67	96	74	65	87	78	89

The teacher would like to use these results for report card grades. One student, however, was absent for the first test, but scored 52% on the second test.

a Justify that a linear regression can be used to estimate this student's result on the first test, and find this value.

b Determine if there is significant correlation at the 5% significance level between the two variables.

a

	ues	yvalues	C	D	E	F
=						=TwoVar('xvalues, 'yvalues, 1):
4	50	30			Σx^2	83465.
5	67	67			sx := sn-...	15.4123
6	00	96			$\sigma x := \sigma n...$	14.7561
7	00	74			n	12.
8	73	65			ȳ	70.75
9	90	87			Σy	849.
10	83	78			Σy^2	63693.
11	94	89			sy := sn-...	18.1565
12	83	78			$\sigma y := \sigma n...$	17.3835
13					Σxy	72266.
14					r	0.837269
15					MinX	50.

C12

	A xvalues	B yvalues	C	D	E	F
=					=LinRegMx('yvalue	
1	65	52		Title..	Linear Regression...	
2	88	57		Reg..	m*x+b	
3	92	76		m	0.71072	
4	50	30		b	31.7999	
5	67	67		r^2	0.701019	
6	100	96		r	0.837269	
7	100	74		Res..	(-3.75732506032...	
8	73	65				
9	90	87				
10	83	78				
11	94	89				
12	83	78				

E1 = "Linear Regression (mx+b)"

$x = 0.7107(52) + 31.8 = 68.8$, hence the first test's result would have been about 69%.

Since r = 0.837, there is a strong positive correlation, hence we can find the line of best fit.

Find the X on Y line by letting Y be the independent variable, and X the dependent.

Substitute for y to find x.

b $H_0: \rho = 0; H_1: \rho \neq 0$

$$t = r\sqrt{\frac{n-2}{1-r^2}} \Rightarrow t = 0.837\sqrt{\frac{12-2}{1-(0.837)^2}} \approx 4.84$$

Step 1: State the null and
alternative hypotheses.
Step 2: Find t.

Step 3: Find the p-value using a
GDC.

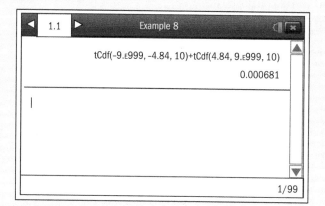

◄ 1.1 ►   Example 8

tCdf(-9.ᴇ999, -4.84, 10)+tCdf(4.84, 9.ᴇ999, 10)

0.000681

1/99

$p = P(T \leq -4.84) + P(T \geq 4.84) = 0.000681$

The p-value 0.000681 is less than 5%, so we reject H_0,
hence there is a significant correlation between the
random variables X and Y.

Step 4: State the conclusion

Alternatively to the above methods,
use the 'LinReg t-test' under the
stats tests menu of your GDC, and
you can see r, t, p-value, and
equation of regression line all in
one screen.

●	A xs	B ys	C	D	E	F
=					=LinRegT	
2	88	57		Alternate...	β & ρ ≠...	
3	83	78		RegEqn	a+b*x	
4	92	76		t	4.8422	
5	50	30		PVal	0.000679	
6	67	67		df	10.	
7	100	96		a	31.7999	
8	100	74		b	0.71072	
9	73	65		s	8.83862	
10	90	87		SESlope...	0.146776	
11	83	78		r²	0.701019	
12	94	89		r	0.837269	

E1	="Linear Reg t Test"

Since p-value $= 0.00679 < 0.05$, we reject H_0.

Exercise 4D

1   The scores of 10 students in a school physical fitness test are shown below, with the higher scores indicating better physical fitness. As shown in the table, the data gathered on the students includes height in cm, weight in kg, and age in years.

Student	S-Score	A-Age	H-Height	W-Weight
1	62	10	145	44.9
2	94	11	150	42.1
3	81	8	139	30.2
4	62	9	148	41.4
5	58	8	130	41.8
6	86	10	150	38.4
7	59	11	154	54.1
8	72	9	140	38.6
9	54	10	153	52.4
10	60	9	120	38.6

A regression line is to be used to predict the physical fitness scores of students by fitting S with one of the variables A, H or W.

a   By finding r for S and A, S and H, and S and W, determine which variable should be used with S to find the equation of the regression line.

b   Find the equation of the regression line.

c   Determine at the 5% level of significance if the correlation between S and the variable you chose is significant.

2   The body mass in grams and the heart mass in mg of ten one-year-old mice are given in the table below.

Body	30	37	38	32	36	32	33	38	44	38
Heart	136	156	150	140	155	157	143	160	170	144

Justify the use of linear regression to estimate the heart mass of a one-year-old mouse whose body mass is 35g, and find this value.

3 The table below shows the blood pressure values of 10 adults in millimetres of mercury, mm Hg. Blood pressure consists of two measurements: systolic (pressure when the heart muscle contracts) and diastolic (pressure when the heart is at rest between beats). Assume that the values form a random sample from a bivariate distribution with correlation coefficient ρ.

X-Systolic	120	125	130	135	140	145	150	155	160	165
Y-Diastolic	80	90	92	98	100	103	105	108	110	110

a State a suitable hypothesis.
b Determine the product moment correlation coefficient for this data and state its p-value.
c Interpret the p-value in the context of the problem.
d Justify finding the line of best fit of Y on X, and use it to estimate the diastolic value of an adult who has a systolic value of 138 mm Hg.

4 An experiment is performed five times by measuring the temperatures needed, Y, to dissolve a certain substance that weighs X grams in a chemical compound. The summary statistics are:

$$\sum x = 182; \quad \sum y = 200; \quad \sum y^2 = 9850; \quad \sum xy = 8390.$$

Find the regression line of X on Y, and use it to find the weight of the substance, x grams, that will dissolve in a litre of the chemical at 90 degrees.

5 The summary statistics for a given data set is as follows:

$$n = 10; \quad \sum x = 30; \quad \sum x^2 = 220; \quad \sum y = 86; \quad \sum y^2 = 1588; \quad \sum xy = 580.$$

Find the equations of the regression lines a Y on X and b X on Y.

Review exercise

EXAM STYLE QUESTIONS

1 It has been determined that the temperature in car tires, X (in °C), varies with the speed of the car, Y (in km/h), according to a linear regression model. The summary statistics are given as:

$$n = 8; \quad \sum x = 440; \quad \sum x^2 = 28400; \quad \sum y = 606; \quad \sum y^2 = 49278; \quad \sum xy = 37000.$$

a Find the correlation coefficient, and determine if there is evidence of significant correlation at the 5% level.
b The speeds in the data vary from 20 km/h to 90 km/h. Using an appropriate regression line, estimate the tire temperature of a car driving at an average speed of 55 km/h.

2 If $\text{Var}(X) = 15$ and $\text{Var}(Y) = 7$ for two random variables X and Y, determine the largest possible covariance of X and Y.

3 Find the smallest r value that will give significant evidence of a positive correlation at the 10% significance level when $n = 20$.

4 The independent random variables X, Y and Z each have a mean of 0 and a variance of σ^2. If $U = X + Y$, $V = X + Z$ and $W = X - Z$, determine the correlation of the following variables and determine if the value found implies that the variables in each case are independent: **a** U and V and **b** V and W.

5 The percentage of a certain drug absorbed by the adult body was tested in ten patients at two different times, and the following data recorded.

1st time	44.5	10.3	20.1	55.0	39.6	24.1	31.2	9.5	22.3	35.1
2nd time	41.2	11.1	18.7	52.3	41.2	26.5	29.3	11.2	25.1	33.2

a Determine the product moment correlation coefficient for this data, and state its p-value.

b Interpret your p-value in the context of the problem and find the equation of the regression line.

c An 11th patient was unable to be tested the 2nd time, but had a drug absorption rate of 19.8 the 1st time. Predict what the absorption rate the 2nd time would have been for this patient.

d The next time, a group of 25 children underwent the same absorption tests for this drug, and it was found that the correlation coefficient between the two different times was 0.532. Use a 1% significance level to determine if there is a positive correlation in the absorption rates between the two different times the drug was administered.

6 a Show that if $r = 0$, the two regression lines, Y on X and X on Y are mutually perpendicular.

b Show that if $r = \pm 1$, the regression lines of Y on X and X on Y are identical.

c Given that the equation of the regression line of Y on X is

$y = a + bx$, where $b = \dfrac{\text{Cov}(X, Y)}{\text{Var}(X)}$, and the regression line of

X on Y is $x = c + dy$, where $d = \dfrac{\text{Cov}(X, Y)}{\text{Var}(Y)}$, show that $r = +\sqrt{bd}$

if b and d are both positive, and $r = -\sqrt{bd}$ if b and d are both negative.

d If the least squares regression line of Y on X is given by $y = 12 + 0.19x$, and the least squares regression line of X on Y is given by $x = -4.4 + 0.77y$, find r, the product correlation coefficient.

Chapter 4 summary

Definition: The observed value r of the sample linear correlation coefficient is defined as

$$r = \frac{\sum_{i=1}^{n}(x_i - \bar{x})(y_i - \bar{y})}{\sqrt{\sum_{i=1}^{n}(x_i - \bar{x})^2 \sum_{i=1}^{n}(y_i - \bar{y})^2}} = \frac{\sum d_x d_y}{\sqrt{\sum d_x^2 \sum d_y^2}}$$

The sample product moment correlation coefficient R, for n paired observations (x, y) on X and Y, is

$$R = \frac{\sum_{i=1}^{n}(X_i - \bar{X})(Y_i - \bar{Y})}{\sqrt{\sum_{i=1}^{n}(X_i - \bar{X})^2 \sum_{i=1}^{n}(Y_i - \bar{Y})^2}}$$

An alternative formula for R is

$$R = \frac{\sum_{i=1}^{n} X_i Y_i - n\bar{X}\bar{Y}}{\sqrt{\sum_{i=1}^{n}(X_i^2 - n\bar{X}^2)(Y_i^2 - n\bar{Y}^2)}}$$

We can classify the strength of the correlation between the random variables X and Y by using the following *general* classification:

- ± 1 indicates a perfect positive/negative correlation.
- $0.5 \leq R < 1$ indicates a strong positive correlation.
- $-1 \leq R \leq -0.5$ indicates a strong negative correlation.
- $0.1 \leq R < 0.5$ indicates a weak positive correlation.
- $-0.5 < R \leq -0.1$ indicates a weak negative correlation.
- $-0.1 < R < 0.1$ indicates a highly weak correlation, or no correlation.

The population product moment coefficient is

$$\rho = \frac{E[(X - E[X])(Y - E[Y])]}{\sqrt{E[(X - E[X])^2] E[Y - E[Y])^2]}}$$

or

$$\rho = \frac{\text{Cov}(X, Y)}{\sqrt{\text{Var}[(X)\,\text{Var}(Y)}}$$

Definition: For each set of n paired observations, the covariance of the random variables X and Y of a joint distribution, $\text{Cov}(X, Y)$, is a measure of the mean value of the product of the two variables' deviations from their respective means, i.e.

$$\text{Cov}(X, Y) = E[(X - E(X))(Y - E(Y))] = E[(X - \mu_x)(Y - \mu_y)],$$

where $\mu_x = E(X)$, $\mu_y = E(Y)$

or, $\text{Cov}(X, Y) = E(XY) - \mu_x \mu_y$

Properties of covariance

1 Symmetry: $\text{Cov}(X, Y) = \text{Cov}(Y, X)$, and $\text{Var}(X) = \text{Cov}(X, X)$

2 For two independent events X and Y, $\text{Cov}(X, Y) = 0$.

Theorem 1: If X and Y are independent variables, then $\rho = 0$.

Theorem 2: If X and Y have a linear relationship, i.e. $Y = mX + c$, then $\rho = \pm 1$.

Theorem 3: For X and Y with a bivariate normal distribution, X and Y are independent if and only if $\rho = 0$.

If X and Y have a bivariate normal distribution such that $\rho = 0$, then the sampling distribution $R\sqrt{\dfrac{n-2}{1-R^2}}$ has the *student's t-distribution* with $(n-2)$ degrees of freedom.

Using the principle of least squares, the Y on X regression line is given by the following formulae:

$$y - \bar{y} = \left(\frac{\sum\limits_{i=1}^{n}[(x_i - \bar{x})(y_i - \bar{y})]}{\sum\limits_{i=1}^{n}(x_i - \bar{x})^2} \right)(x - \bar{x}) \text{ or } y - \bar{y} = \left(\frac{\sum\limits_{i=1}^{n}x_i y_i - n\bar{x}\,\bar{y}}{\sum\limits_{i=1}^{n}x_i^2 - n\bar{x}^2} \right)(x - \bar{x})$$

The gradient of the Y on X regression line is $\dfrac{\text{Cov}(X,Y)}{\text{Var}(X)}$

The formula for the X on Y regression line is:

$$x - \bar{x} = \left(\frac{\sum\limits_{i=1}^{n}(x_i - \bar{x})(y_i - \bar{y})}{\sum\limits_{i=1}^{n}(y_i - \bar{y})^2} \right)(y - \bar{y}) = \left(\frac{\sum\limits_{i=1}^{n}x_i y_i - n\bar{x}\,\bar{y}}{\sum\limits_{i=1}^{n}y_i^2 - n\bar{y}^2} \right)(y - \bar{y})$$

The gradient of the X on Y regression line is $\dfrac{\text{Cov}(X,Y)}{\text{Var}(Y)}$

Answers

Chapter 1

Skills check

1 a $\text{Mode}(X) = -1, 1$
   Median, $m = 1$
   $\mu = 0.95$
   $\sigma = 1.72$
b $\text{Mode}(X) = 1$
   Median, $m = 2$
   $\mu = 2$
   $\sigma = 1$

2 a $\text{Mode}(X) = 0$
   Median, $m = 0.695$
   $\mu = \dfrac{3}{4}$
   $\sigma = 0.487$
b $\text{Mode}(X) = 0$
   Median, $m = 0$
   $\mu = 0$
   $\sigma = 0.342$

c $\text{Mode}(X) = 3$
   Median, $m = 4$
   $\mu = 4.16$
   $\sigma = 0.839$

3 a $\dfrac{2}{3}$ **b** $2 + 2\sqrt{2}$

4 a $f'(x) = \dfrac{1}{(2-x)^2}, x \neq 2;$
   $\displaystyle\int f(x)dx = -\ln(2 - x) + c, x < 2$
b $f'(x) = 3e^{3x+1}; \displaystyle\int f(x)dx = \dfrac{1}{3}e^{3x+1} + c$
c $f'(x) = -\cos\left(\dfrac{\pi - 2x}{3}\right);$
   $\displaystyle\int f(x)dx = \dfrac{9}{4}\cos\left(\dfrac{\pi - 2x}{3}\right) + c$
d $f'(x) = 4x(x^2 - 2); \displaystyle\int f(x)dx = \dfrac{x^5}{5} - \dfrac{4}{3}x^3 + 4x + c$

Exercise 1A

1 a $k = 3$ **b** $F(x) = \begin{cases} 0, & x < 0 \\ \dfrac{x^2 + 7x + 6}{24}, & x = 0, 1, 2 \\ 1, & x > 2 \end{cases}$

2 a $a = 10$
b $F(x) = \begin{cases} 0, & x < 1 \\ \dfrac{9x - x^2}{20}, & x = 1, 2, 3, 4 \\ 1, & x > 4 \end{cases}$
c $P(X \le 2) = \dfrac{7}{10}$

3 a $P(X = x) = \begin{cases} \dfrac{x+1}{6}, & x = 0, 1, 2 \\ 0, & \text{otherwise} \end{cases}$
b $\text{Mode} = 2$

4 a $P(X = x) = \begin{cases} \dfrac{x}{25}, & x = 1, 3, 5, 7, 9 \\ 0, & \text{otherwise} \end{cases}$ **b** $m = 7$

5 a $b = 1$
b $F(x) = \begin{cases} 0, & x < 0 \\ 2 - 2\cos x, & 0 \le x \le \dfrac{\pi}{3} \\ 1, & x > \dfrac{\pi}{3} \end{cases}$
c $P\left(X \ge \dfrac{\pi}{6}\right) = 0.732$

6 a The function is well-defined if $\displaystyle\int_{-\infty}^{+\infty} f(x)dx = 1$
b The modal value doesn't exist.

Exercise 1B

1 a $P(X = 2) = 0.24$ **b** $P(X = 3) = 0.104$
   c $P(X = 4) = 0.0625$ **d** $P(X = 5) = 0.000182$
2 a $P(X \le 4) = 0.684$ **b** $P(X > 6) = 0.000729$
   c $P(5 \le X \le 7) = 0.158$ **d** $P(1 < X \le 7) = 0.009$
3 $P(X \le 3) = 0.980$
4 a $P(X = 5) = 0.0000377$ **b** $P(X \ge 4) = 0.000512$
5 a $P(X = 4) = 0.0921$ **b** $P(X \ge 7) = 0.377$

Investigation

   a 0.36 **b** 0.0081 **c** 0.409

Exercise 1C

1 For 1B, Question 1:
   a $E(X) = 1.67, \text{Var}(X) = 1.11$
   b $E(X) = 7.14, \text{Var}(X) = 43.9$
   c $E(X) = 2, \text{Var}(X) = 2$
   d $E(X) = 1.14, \text{Var}(X) = 0.155$

   For 1B, Question 2:
   a $E(X) = 4, \text{Var}(X) = 12$
   b $E(X) = 1.43, \text{Var}(X) = 0.612$
   c $E(X) = 3.33, \text{Var}(X) = 7.78$
   d $E(X) = 1.01, \text{Var}(X) = 0.00916$
2 a $E(X) = 1.37$
   b Mario must make four shots to destroy the balloon.
3 6 students

Exercise 1D

1 a $P(X = 2) = 0.16$ **b** $P(X = 4) = 0.188$
   c $P(X = 9) = 0.235$ **d** $P(X = 32) = 0.0847$
2 a $P(X \le 4) = 0.508$ **b** $P(X > 6) = 0.109$
   c $P(5 \le X \le 7) = 0.525$ **d** $P(8 < X \le 11) = 0.326$
3 a $p = 0.4$ **b** $P(3 \le X \le 5) = 0.503$
4 a $X \sim \text{NB}\left(2, \dfrac{1}{4}\right)$ **b** 3 **c** $\dfrac{47}{128}$
5 a 0.313 **b** 0.0473
6 a 0.264
   b $0.999999 \approx 1$, so it is almost certain that he will not need to interview more than a dozen students.
7 a 0.0829 **b** 0.589

Exercise 1E

1 $G(t) = \dfrac{1}{16} + \dfrac{1}{4}t + \dfrac{3}{8}t^2 + \dfrac{1}{4}t^3 + \dfrac{1}{16}t^4$

x_i	0	1	2	3	4
p_i	$\dfrac{1}{16}$	$\dfrac{4}{16}$	$\dfrac{6}{16}$	$\dfrac{4}{16}$	$\dfrac{1}{16}$

2

x_i	1	2	3	...	k	...
p_i	$\dfrac{1}{6}$	$\dfrac{5}{36}$	$\dfrac{25}{216}$	...	$\underbrace{\left(\dfrac{5}{6}\right)^{k-1}}_{\text{not one}} \times \underbrace{\dfrac{1}{6}}_{\text{one}} = \dfrac{5^{k-1}}{6^k}$	...

156   Answers

3 a $P(X = 0) = \dfrac{2}{3}$ **b** $P(X \le 1) = \dfrac{8}{9}$

c $P(X \ge 3) = \dfrac{1}{27}$ **d** $\dfrac{1}{3^k}$

4 a $E(X) = p$, $Var(X) = p(1-p)[= pq]$

b $E(X) = \dfrac{r}{p}$, $Var(X) = \dfrac{rq}{p^2}$

5 For question 1   For question 2
$E(X) = 2$, $Var(X) = 1$ $E(X) = 6$, $Var(X) = 30$
For question 3
$E(X) = \dfrac{1}{2}$, $Var(X) = \dfrac{3}{4}$

6 a $P(X = 1) = \dfrac{4}{7}$ **b** $G(t) = \dfrac{4t + 2t^2}{7 - t^2}$

c The expected number of shots is 2.

d The maximum number of shots is 5.

Exercise 1F

1 a $G_{X+Y}(t) = \left(\dfrac{2 + 7t + 3t^2}{12}\right)^2$

b $P(X + Y \le 1) = \dfrac{2}{9}$ **c** $E(X + Y) = \dfrac{13}{6}$

2 a $G_{X+Y}(t) = \left(\dfrac{t^2}{6 - 7t + 2t^2}\right)^3$

b $E(X + Y) = 15$ **c** $Var(X + Y) = 24$

3 a $G_X(t) = e^{0.25(t-1)}$, $G_Y(t) = e^{0.15(t-1)}$, $G_Z(t) = e^{0.05(t-1)}$

b 0.989

4 $\left(\dfrac{pt}{1 - qt}\right)^r$

Review exercise

1 a $P(1 \le X \le 4) = \dfrac{624}{625}$ **b** $P(X \ge 2) = \dfrac{1}{25}$

c $E(X) = \dfrac{5}{4}$ **d** $Var(X) = \dfrac{5}{16}$

2 a $G_X(t) = e^{0.6(t-1)}$, $G_Y(t) = e^{0.12(t-1)}$, $G_Z(t) = e^{0.28(t-1)}$

b 1 **c** 0.264

3 a i 0.0117 **ii** 0.105 **iii** 0.0625 **iv** 0.922

b Eight

4 a $a = 2$ **b** $f(x) = \begin{cases} \dfrac{x}{2}, & 0 \le x \le z \\ 0, & \text{otherwise} \end{cases}$ **c** 2

5 b $E(X) = a$

Chapter 2

Skills check

1 a $P(X = 2) = 0.311$ **b** $P(1 \le X \le 3) = 0.786$

2 $a_1 = -\dfrac{1}{2}$, $a_2 = -\dfrac{1}{3}$, $a_3 = 2$

Exercise 2A

1 a $E(3X) = 15.9$, $Var(3X) = 10.8$

b $E(X + 3) = 8.3$, $Var(X + 3) = 1.2$

c $E(4X + 1) = 22.2$, $Var(4X + 1) = 19.6$

d $E(2X - 5) = 5.6$, $Var(2X - 5) = 4.8$

e $E(kX + p) = 5.3k + p$, $Var(kX + p) = 1.2k^2$

2 a $E(3X + 2) = 14$ **b** $Var(3X - 2) = 21.6$

3 $E(2Y - 1) = 2$, $Var(2Y - 1) = 3$

4 a $E(3 - 2Y) = -1$ **b** $Var(3 - 2Y) = 8$

5 a $E(2X - 3) = 45$ **b** $Var(2X - 11) = 192$

6 $Var(5X + 3) = 90$

7 $E(3X + 2) = 2\sqrt{6} + 2$, $Var(3X + 2) = 3$

Exercise 2B

1 a $E(X + Y) = -2$, $Var(X + Y) = 1.9$

b $E(2Y - Z) = -22$, $Var(2Y - Z) = 8.4$

c $E(2Z - 7X) = 3$, $Var(2Z - 7X) = 35.7$

d $E(X - Y + Z) = 20$, $Var(X - Y + Z) = 4.7$

e $E(X + Y - Z) = -14$, $Var(X + Y - Z) = 4.7$

f $E(3Z - 2X + 4Y) = 10$
$Var(3Z - 2X + 4Y) = 49.6$

2 a $E(3X + 5Y) = 31$ **b** $Var(11Y - 7X) = 703$

3 $Var(2X - 3Y) = 72 - 72p$

4 $Var(X - Y) = \dfrac{20}{p} - 20$

Exercise 2C

1 a

$X = x$	0	1
$P\{X = x\}$	$\dfrac{1}{2}$	$\dfrac{1}{2}$

$E(X) = \dfrac{1}{2}$, $Var(X) = \dfrac{1}{4}$

b Since we have 6 independent flips of the same coin, we add six instances of the variable X.

c $E(Y) = 3$, $Var(Y) = \dfrac{3}{2}$ **d** $y \in [-0.67, 6.67]$

2 a

$X = x$	0	1
$P\{X = x\}$	$\dfrac{2}{3}$	$\dfrac{1}{3}$

$E(X) = \dfrac{1}{3}$, $Var(X) = \dfrac{2}{9}$

b Since we have 4 independent rolls of the same die, we add four instances of the variable X.

c $E(Y) = \dfrac{4}{3}$, $Var(Y) = \dfrac{8}{9}$

3 a $E(X + X + X) = 9$, $Var(X + X + X) = 12$

b $E(3X) = 9$, $Var(3X) = 36$

c $Var(X + X + X + 3X) = 48$
$Var(6X) = 144$

4 a $E(X + X + X + X + X) = 10$
$Var(X + X + X + X + X) = 5$

b $E(Y + Y + Y) = 15$ $Var(Y + Y + Y) = 9$

c $E(X + X + X + X + X + Y + Y + Y) = 25$
$Var(X + X + X + X + X + Y + Y + Y) = 14$

5 a

$X = x_i$	1	2	3	4
$P\{X = x_i\}$	$\dfrac{1}{4}$	$\dfrac{1}{4}$	$\dfrac{1}{4}$	$\dfrac{1}{4}$

$E(X) = \dfrac{5}{2}$

$Y = y_i$	1	2	3
$P\{Y = y_i\}$	$\dfrac{1}{2}$	$\dfrac{1}{3}$	$\dfrac{1}{6}$

$E(Y) = \dfrac{5}{3}$

b

$Z = z_i$	1	2	3	4	6	8	9	12
$P\{Z = z_i\}$	$\frac{1}{8}$	$\frac{5}{24}$	$\frac{1}{6}$	$\frac{5}{24}$	$\frac{1}{8}$	$\frac{1}{12}$	$\frac{1}{24}$	$\frac{1}{24}$

$$E(Z) = \frac{25}{6}$$

6 $\mu = 2$

Exercise 2D

1 a $P(Y - Z - W < 0) = 0.5$
  b $P(X + Y + Z + W > 0) = 0.892$
  c $P(3X + Y > Z + W) = 0.5$
  d $P(X - 3Z \leq 2Y + W) = 0.329$
  e $P(X - 4Z \leq 3X - 2Z) = 0.151$
  f $P(W - Y \leq 2Y + 3W) = 0.994$
2 a 0.551   **b** 0.162
3 a 0.159   **b** 0.201   **c** 0.564
4 a 0.0186   **b** 0.0000155   **c** 0.149
  d 0.118

Exercise 2E

1 a $P(1 \leq \bar{X} \leq 3) = 0.777$
  b $P(2 \leq \bar{X} \leq 8) = 0.992$
  c $P(|\bar{X}| \geq 0.8) = 0.639$
2 0.053
3 a $P(\bar{X} \geq 37) = 0.228$
  b $P(X + X + X + X + X > 180) = 0.355$
4 a $P(\bar{X} \leq 70) = 0.920$ **b** $P(T \leq 800) = 0.782$

Exercise 2F

1 a $P(1.5 \leq \bar{X} \leq 2.5) = 0.639$
  b $P(1.25 \leq \bar{X} \leq 1.35) = 0.923$
  c $P(\bar{X} \geq -0.48) = 0.421$
  d $P(\bar{X} < 397) = 0.0353$
  e $P\left(-\frac{1}{2} < \bar{X} < \frac{1}{2}\right) = 0.0983$
  f $P(1.8 < \bar{X} < 2.2) = 0.280$
2 0.834
3 a 0.193   **b** 0.00469   **c** 0.917
4 a 0.579   **b** 62

Review exercise

1 0.796
2 b $P(X \geq 6) = 0.384$ **c** $P(X + Y < 5) = 0.173$
  d $E(Z) = 7$
    $Var(Z) = 77$
  e The random variable Z has no Poisson
    distribution since $7 = E(Z) \neq Var(Z) = 77$
3 a 0.00621   **b** 0.0787
  c $N(720, 20^2 + 12^2) = N(720, (4\sqrt{34})^2)$
  d 0.299
4 a 20 or 47   **b** 16
5 a 0.377   **b** 0.196   **c** 45

Chapter 3

Skills check

1 $\bar{x} = 3.62$, $\sigma^2 = 2.82$
2 a $P(X = 5) = 0.03125$ **b** $P(3 \leq X < 8) = 0.322$
3 a $P(Y = 0) = 0.670$ **b** $P(3 \leq X < 8) = 0.569$

Exercise 3A

1 a $\bar{x} = 11$, $s^2 = 33$ **b** $\bar{x} = 29.75$, $s^2 = 25.3$
  c $\bar{x} = 67$, $s^2 = 1518$
2 $\bar{x} = 0.65$, $s^2 = 0.864$
3 a $\bar{x} = 33.7$ **b** $s^2 = 23.8$

Exercise 3B

1 a [4.19, 5.81]   **b** [−12.8, −9.20]
  c [2819, 2889]
2 a [3.90, 6.10]   **b** [0.104, 0.167]
  c [320, 325]
3 $n = 30$
4 [72.3 g, 77.2 g]
5 a $\bar{x} = 15.8$ **b** $n = 10$

Investigation

a i [95.0, 105.0]   **ii** [96.9, 103.1]
  iii [97.8, 102.2]   **iv** [98.7, 101.3]
b At the same significance level, the larger the sample
  size we take, the narrower the confidence interval
  we get.

Exercise 3C

1 a [14.45, 15.55]   **b** [−25.81, −20.19]
  c [3430, 3526]
2 a [1.94, 8.06]   **b** [0.112, 0.159]
  c [319.6, 324.9]
3 [67.6 g, 82.5 g]
4 a $\bar{x} = 13.35$ **b** The confidence level is 90%.
5 a [483.4, 588.6]   **b** [463.0, 609.0]
  c For the same set of data, a higher significance
    level will mean a wider confidence interval.

Exercise 3D

1 a [−2.18, 1.98]   **b** [−13.73, −7.44]
  c [−0.0609, 0.0859]
2 a

$d_i =$ Bob−Rick	3	−8	−3	9	6	−9	13	2	−4	−3	3	−7

  b [−4.27, 4.60]

Exercise 3E

1 a Since the p-value is $0.000008 < 0.1$ we reject the
  null hypothesis at the 10% significance level.
  b Since the p-value is $0.095581 > 0.05$ we have no
    sufficient evidence to reject the null hypothesis at
    the 5% significance level.
  c Since the p-value is $0.010566 > 0.01$ we have no
    sufficient evidence to reject the null hypothesis
    at the 1% significance level.

2 a Since the p-value is $0.131776 > 0.1$ we have no sufficient evidence to reject the null hypothesis at the 10% significance level.

b Since the p-value is $0.029673 > 0.01$ we have no sufficient evidence to reject the null hypothesis at the 1% significance level.

c Since the p-value is $0.005283 < 0.01$ we reject the null hypothesis at the 1% significance level.

3 a Since the p-value is $0.036819 < 0.05$ we reject the null hypothesis at the 5% significance level.

b Since the p-value is $0.109391 > 0.1$ we have no sufficient evidence to reject the null hypothesis at the 10% significance level.

c Since the p-value is $0.000153 < 0.01$ we reject the null hypothesis at the 1% significance level.

4 a H_0: "The mean weight is 26 g." ($\mu = 26$)
H_1: "The mean weight is not 26 g." ($\mu \neq 26$)

b Use the z-test. Since the p-value is $0.00001 < 0.01$ we reject the null hypothesis at the 1% significance level and conclude that the harvested snails are not from the population.

5 a H_0: "The mean level of fat in the drink is 1.4 g." ($\mu = 1.4$)
H_1: "The mean level of fat in the drink is more than 1.4 g." ($\mu > 1.4$)

b Use the z-test. Since the p-value is $0.335687 > 0.05$ we have no sufficient evidence to reject the null hypothesis at the 5% significance level and conclude that the company's claim is correct.

6 a H_0: "The mean volume of juice in the bottle is 300 ml." ($\mu = 300$)
H_1: "The mean volume of juice in the bottle is less than 300 ml." ($\mu < 300$)

b Use the z-test. Since the p-value is $0.004612 < 0.1$ we reject the null hypothesis at the 10% significance level and conclude that the bottles contain less volume than stated.

Exercise 3F

1 a Since the p-value $0.6382 > 0.05$ we have no sufficient evidence to reject the null hypothesis at the 5% significance level.

b Since the p-value $0.10858 > 0.1$ we have no sufficient evidence to reject the null hypothesis at the 10% significance level.

c Since the p-value $0.013122 > 0.01$ we have no sufficient evidence to reject the null hypothesis at the 1% significance level.

2 a Since the p-value $0.027947 < 0.05$ we reject the null hypothesis at the 5% significance level.

b Since the p-value $0.007494 < 0.01$ we reject the null hypothesis at the 1% significance level.

c Since the p-value $0.225675 > 0.1$ we have no sufficient evidence to reject the null hypothesis at the 10% significance level.

3 a Since the p-value $0.743755 > 0.1$ we have no sufficient evidence to reject the null hypothesis at the 10% significance level.

b Since the p-value is $0.004763 < 0.05$ we reject the null hypothesis at the 5% significance level.

c Since the p-value $0.115078 > 0.01$ we have no sufficient evidence to reject the null hypothesis at the 1% significance level.

4 H_0: "The mean volume is 120 ml." ($\mu = 120$)
H_1: "The mean volume is not 120 ml." ($\mu \neq 120$)
Since the p-value $0.283654 > 0.01$ we have no sufficient evidence to reject the null hypothesis at the 1% significance level and conclude that the factory advertised a correct volume of a particular ice-cream product.

5 H_0: "The mean life expectancy is 30,000 hours." ($\mu = 30,000$)
H_1: "The mean life expectancy is less than 30,000 hours." ($\mu < 30,000$)
Since the p-value $0.094543 < 0.1$ we reject the null hypothesis at the 10% significance level and conclude that the manufacturer claims a longer life expectancy of the LED lamps.

Exercise 3G

1 H_0: "There is no difference in finishing times." ($\mu_d = 0$)
H_1: "There is a difference in finishing times." ($\mu_d \neq 0$)
Use the t-test. Since the p-value $0.874063 > 0.05$ we have no sufficient evidence to reject the null hypothesis at the 5% significance level and conclude that there is no difference in finishing times on the two Rubik's Cubes.

2 H_0: "There is no difference in the scores." ($\mu_d = 0$)
H_1: "There is a difference in the scores." ($\mu_d \neq 0$)
Use the t-test. Since the p-value $0.085622 < 0.1$ we reject the null hypothesis at the 10% significance level and conclude that players score a better result when using the new type of dart.

3 H_0: "There is no difference in the weights." ($\mu_d = 0$)
H_1: "Students who join the programme drop some weight." ($\mu_d > 0$)
Use the t-test. Since the p-value $0.121576 > 0.05$ we have no sufficient evidence to reject the null hypothesis at the 5% significance level and conclude that there is no difference in weight before and after the programme.

Exercise 3H

1 a 0.0455   **b** 0.773
2 a $E(X) = 25$; $\alpha = 0.00130$ **b** 0.978
3 a 0.133   **b** 0.972

Review exercise

1 a [602, 702]   **b** [586, 718]
c We notice that a higher significance level means a wider confidence interval.

2 a

Difference	-2	9	-6	8	5	-4	15	-8	7	-5

b H_0: "There is no difference in potassium levels." ($\mu_d = 0$)

H_1: "There is a difference in potassium levels." ($\mu_d \neq 0$)

Since the p-value $0.46297 > 0.01$ we have no sufficient evidence to reject the null hypothesis at the 1% significance level and we conclude that there is no difference in measurement of the two types of biochemical analyzers.

3 a $\bar{x} = 5.33$ $s^2 = 4.38$

b [3.72, 6.94]

c In 99% of the cases, the mean value of a sample of 15 observations taken from the population will fall within the confidence interval [3.72, 6.94].

4 a $\bar{x} = 51.2$ **b** 95%

5 a $30\,s$ **b** $\bar{x} = 30.97$ $s^2 = 0.298$

c H_0: "The average time is 30 s." ($\mu = 30$)

H_1: "The average time is more than 30 s." ($\mu > 30$)

We use t-test since the standard deviation is unknown.

Since the p-value $0.000163 < 0.05$ we reject the null hypothesis and conclude that the average time is more than 30 s, therefore the company sets up the speedometers to show a higher speed.

6 a $\bar{x} = 210$

b $n = 15.37$, thus the sample size should be 16.

7 a i $\bar{x} = 23.5$ **ii** $s = 10.5$

b i [21.8, 25.2] **ii** [22.0, 24.9]

c $[22.0, 24.9] \subset [21.8, 25.2]$, so a 90% confidence interval is a subset of a 95% confidence interval.

8 a i $]-\infty, 9.19[$ **ii** $]-\infty, 8.96[$

b i 0.569 **ii** 0.705

c When the probability of a Type I error decreases from 10% to 5%, the probability of a Type II error increases from 0.569 to 0.705.

Chapter 4

Skills check

1 a $2E(Z) - 3E(Y) + 2E(X)$

b $4Var(Z) - 9Var(Y) + 4Var(X)$

c $E(X)E(Y)E(Z)$

2 a $\bar{x} = 43.5$, $\bar{y} = -31.75$,

$Var(X) = 46.25$, $Var(Y) = 98.69$

3 a 0.382 **b** 0.951 **c** 0.938 **d** 0.732

Exercise 4A

1 $r = -0.382$; there is a weak negative correlation

2 $r = 0.794$; there is a strong positive correlation

3 $r = -0.970$; there is a strong negative correlation

Exercise 4C

1 $p = 0.00229$. Since $p < 0.01$, we reject the null hypothesis. There is evidence of significant correlation between the two variables at the 1% level.

2 $p = 0.394$. Since $p > 0.05$, there is not enough evidence to reject the null hypothesis, i.e. there is not evidence of significant correlation between the two variables at the 5% level.

3 $p = 0.0343$. Since $p < 0.1$, we reject the null hypothesis. There is evidence of significant correlation between the two variables at the 10% level.

Exercise 4D

1 a Use W, because it has the strongest correlation, $r = -0.546$

b $S = -0.270W + 61.4$

c $p = 0.102$. Since $p > 0.05$, there is no evidence to reject the null hypothesis, i.e. there is not evidence of significant correlation between the two variables at the 5% level.

2 $r = 0.756$; $y = 1.91x + 82.8$; 150 mg

3 a H_0: $\rho = 0$; H_1: $\rho \neq 0$

b $r = 0.962$; $p = 0.000008$

c There is very strong evidence to indicate a positive association between the random variables X and Y.

d $y = 0.623x + 10.8$; y is approximately 97.

4 $x = 0.6y + 12.4$; 66.4 grams

5 a Y on X: $y = 2.48x + 1.17$

b X on Y: $x = 0.380y - 29.6$

Review exercise

1 a $r = 0.975$, $p = 0.000039$; $p < 0.05$, hence there is sufficient evidence of a strong positive relationship between the two random variables.

b 32.4 °C

2 10.2

3 $r = 0.299$

4 a $\frac{1}{2}$, no **b** 0, no

5 a $r = 0.991$; $p = 2.78 \times 10^{-8}$

b The p-value suggests a strong relationship between the two random variables; $y = 0.905x + 2.59$

c 20.5

d $p = 0.00620$; $p < 0.01$, hence there is a strong correlation between the random variables.

6 d $r = 0.382$

Index

Page numbers in *italics* indicate the Answers section.